China Being Led and Leading

Zhimin Li · Daniel Braun

China Being Led and Leading

A Literary and Cultural Interpretation

palgrave
macmillan

Zhimin Li
Guangzhou University
Guangzhou, China

Daniel Braun
Guangzhou University
Guangzhou, China

ISBN 978-981-96-6802-1 ISBN 978-981-96-6803-8 (eBook)
https://doi.org/10.1007/978-981-96-6803-8

Cover credit: gremlin

This Palgrave Macmillan imprint is published by the registered company Springer Nature Singapore Pte Ltd.
The registered company address is: 152 Beach Road, #21-01/04 Gateway East, Singapore 189721, Singapore

For a ***more*** *beautiful world*
of ***more*** *self-cultivation*
more *mutual understanding*
more *kindness*
and ***more*** *love!*

You and Me,—We are Family

You and me
From same spirit of universe
We are family!

You and me
Through the same evolution
We are family!

You and me
In the same flesh and blood
We are family!

You and me
Of the same heart and mind
We are family!

You and me
Raising same children of humanity
We are family!

Let's come together
Put your hands in mine
We are family!

—Adapted from Beijing 2008 Olympic
Games Theme Song

Acknowledgements Chapter 1 has been published as: The One Way Model of Cultural Interaction: Literary Interactions between China and Cambridge, *The Cambridge Quarterly*, March of 2012.

Chapter 3 has been published as: New Chinese Poetry: The Origin and the Development—From the Perspective of Cultural Exchanges between China and the West, *The International Literary Quarterly*, 2010 (1).

Chapter 4 has been published as: The Interplay of "Secular Morality" and "Philosophy-Science", *NBU Journal of Philosophy*, March of 2011.

Chapter 9 has been published in the online American journal *b20* (8/13/22) as: The Linking Matters: An International Poetics of Sense-Making and Innovation.

All previously published material has been modified for the purposes of this book.

Our sincere thanks go to Prof. Neil Roberts who kindly proofread the whole manuscript and provided many wonderful comments.

Competing Interests The authors have no competing interests to declare that are relevant to the content of this manuscript.

Contents

1 **Preface: A Personal History of Modern Transformation of China** ... 1
Works Cited ... 11

2 **The One-Way Model of Cultural Interaction: Literary Interactions Between China and Cambridge** ... 13
A Brief Review of Interactions Between China and the West ... 13
The One-Way Model of Literary Interaction ... 16
Xu Zhimo and William Empson Crossing the Boundaries ... 18
Conclusion: Balance in Becoming ... 22
Works Cited ... 25

3 **Turbulent Transformation of the Literary and Cultural Mentality of China from 1917 to 1949 and the Vigorous One after the Reform and Opening-Up Policy** ... 27
Endless War and Suffering in China before 1949 ... 27
Literary and Cultural Movements in China from 1917 to 1949 ... 30
  Linguistic Transformation from Wenyan to the Vernacular ... 31
  Transformation of Literary Forms from the Traditional to the Modern ... 33
  Transformation of Cultural Values ... 35
Vigorous Transformation of China after the Reform and Opening-Up Policy ... 37
  From an Agricultural Nation to an Industrial One ... 38
  Modern Values Getting Rooted in China ... 39
Conclusion ... 40
Works Cited ... 40

4 The Origin and Development of New Chinese Poetry from the Perspective of Cultural Exchanges between China and the West 43
Influence of China on the West 43
New Chinese Poetry Under the Influence of the West 45
The Formation Period of New Chinese Poetry from the 1910s to the 1940s 48
The Misty Poetry Period from 1978 to 1989 and the Individual Creation Period after 1990s 51
Conclusion: New Chinese Poetry as a Melting Pot for Traditional Chinese Poetry and Western Poetry 54
Works Cited 57

5 The Interplay of "Common Morality" and "Philosophy-Science" 59
Confucianism as a System of Common Moralities 59
Christianity as a System of Common Moralities 61
Reformation of Christianity 64
The Revival of the Theory of Seeking for Truth 68
Conclusion 70
Works Cited 72

6 Re-examination of the Question "Where Do I Come From?": Establishment of Modern Cultural Norms Related to Parent–Child Relationships 75
Interpretation of the Origin of Humanity and Deduction of Rights and Obligations Related to Parent–Child Relationships in Traditional Chinese and Western Cultures 76
Exploration of the Question "Where Do I Come From?" by Modern Science: Deconstruction of Mythic or Religious Stories and Reaffirmation of Human Divinity 79
Reexamination of the Question of "Where Do I Come From?" and the Resulting Cultural Rights and Obligations in the Modern Age 83
Conclusion 86
Works Cited 87

7 "Where Am I Going?": Reconfirmation of the Fundamental Cultural Faith in Life of Humanity and Its Survival 89
Answering the Question of "Where Am I Going?" in China and the West in the Traditional Age and the Cultural Faith in the Life of Humanity 89
Modern Science's Answering to the Question of "Where Am I Going?" and the Challenges It Poses 92

Answering the Question of "Where Am I Going?" in the Modern Context and the Fundamental Cultural Faith in Life of Humanity and Its Survival Reconfirmed 95
Conclusion: Future-Orientated Cultural Faith and Values in the Modern Age 98
Works Cited 99

8 "Who Am I?": Establishment of Modern Culture on Basis of Three Forms of Personal Life 101
"The Physical Self," "The Social Self" and "The Conceptual Self" 101
Analyses of Problems in Traditional China from the Perspective of the Three Selves 106
Establishment of the Modern Cultural System for Welfare of the Three Selves 109
Conclusion 112
Works Cited 113

9 Body Thinking, Image Thinking and Concept Thinking: The Philosophy of Thinking and Its Implication for Communications 115
Information Process and Body Thinking 116
Image-Encoding and Rules of Thinking 120
Information Connection and Knowledge 123
Conclusion 125
Works Cited 126

10 The Linking Matters: An International Poetics of Sense-Making and Innovation 129
Works Cited 137

11 Truth as a Construct and the Growth of Knowledge 139
Defining Construct-Truth 140
Why is There Construct-Truth? 143
How to Promote the Development of Construct-Truth and the Growth of Knowledge 145
Conclusion 148

12 Survival First: A Review of Modern Humanity and a Prospect for Its Future 151
The Drives for Modernisation, the Problems Caused and the Way Out 152
Survival First: The First Law for Humanity in both the Traditional and Modern ages 155

The Root Faith of Life and the Modern Cultural Value System Based upon It ... 159
Conclusion: A Prospect for the Maturated Modern age of Humanity ... 161
Works Cited ... 162

Coda For the Mature Modern Age ... 163

Index ... 165

Chapter 1
Preface: A Personal History of Modern Transformation of China

I am so fortunate to have witnessed one of the greatest phases of modern transformation of China up till now. I am grateful to so many people who have contributed to this great transformation, both at home and abroad. I am proud that I have myself also done my part, though mine is so humble as a teacher of English, a scholar and a lover of poetry.

I shall start my personal history from my grandpa who died when my father was only four or five years old. I have been trying to figure out what happened to my grandpa from the bits of information I have heard here and there. My grandpa was a bricklayer, a profession that is still taken by a dozen or so people in my home village. He was working at a site at some distance from home when he fell ill and demonstrated symptoms of hematemesis. He was sent home by his fellow villagers, in accordance with a custom that has been practiced in China for thousands of years that a dying person is to be carried back and to die at home. Nobody ever told me what illness it was. However, from the symptoms described, it must have been tuberculosis – a disease that was common in China those days and killed many.

Given the age of my father, I guess my grandpa was in his thirties when he died. It was not big news for people to die so young in those days, as the average lifespan of people in China was around 35 years before 1949. (Lu, 2019) In fact, three people most close to me of the last generation became orphans very young. Like my father, my mother lost her father in her childhood. Her mother remarried, bringing her to a new household. My PhD Director who guided me to an academic life lost his mother very early, and was brought up by his aunt in poverty. In the year of 2020, the average lifespan of people in China had reached around 78, something that could never have been imagined by the generation of my grandpa.[1]

It must have been devastating to my father and the whole family when my grandpa passed away. Soon after, my grandma remarried, leaving my father and his older brother to themselves. I could not imagine how the two small boys even survived.

[1] The exact number is 77.93. See the website of *National Bureau of Statistics* of China: https://data.stats.gov.cn/easyquery.htm?cn=C01&zb=A0305&sj=2000.

Z. Li and D. Braun, *China Being Led and Leading*,
https://doi.org/10.1007/978-981-96-6803-8_1

Later on, a wanderer, quite senior in age, happened to visit our village and somehow decided to settle in my father's shabby house, taking responsibility to look after the two orphans.

This Grandpa Wanderer (may he rest in peace) not only helped save my father from starvation, but actually saved his life in a terrible accident. One day, a villager was cutting small branches on a tall paper mulberry tree with a sickle. My father, a small hungry boy, was standing right below the tree, waiting to pick up and eat some paper mulberry seeds on the branches as they were cut. When the paper mulberry seeds are ripe, they turn red and sweet. I myself ate them in my childhood. Suddenly, the sickle slipped out of the villager's hand and fell right onto my father head, cutting deep into it.

Everybody witnessing it was shocked! The sickle cut so deep into my father's head that no body dared to move it. My father fainted. Grandpa Wanderer got the news and rushed to my father and carried him home. It was said that Grandpa Wanderer was a pious practitioner of a kind of folk religion. He had a table where statues of gods were worshiped with incense sticks often lit. Grandpa Wanderer put my unconscious father on the table, and started to play a kind of ritual during which it was said a certain god would descend from heaven and enter in the person who practiced it. I do not believe the theory. However, I respect all religions because it is good sometimes to believe in them even though the stories they tell or theories they hold might not be provable. Grandpa Wanderer surely believed that a god had entered his body before he grew so brave and removed the sickle from my father's head in an instant, with a power he could not normally have demonstrated. Meanwhile, he grabbed a big handful of ashes from the incense burner and firmly pressed them against the huge cut where blood was gushing out and bound it tightly with a piece of cloth.

It is a miracle that my father survived. My father is over eighty years old now. Nevertheless, a huge scar is still clearly seen right on the top of his head.

I could never have come into being if my father had died in this terrible accident. I often feel that life is a miracle when I am pondering how close my ancestors, including my father, were to death in all sorts of accidents or hard times, and how many times such stories were repeated. There were so many disasters in the recorded history of China, of which some killed even more than half of the whole national population. (Ge 2008, 94)[2] In any of such disasters, one of my ancestors could have died an early death, cutting the heritage chain that connects past to future, myself to my family. I wonder how many times they escaped from death because of a hand given, like that of Grandpa Wanderer. Every time when I meditate on this topic, I realize how much I owe my life to the society I live in and the people who support me, and it seems I hear a call to urge me to do good for society and to treat my fellow people kindly.

When I read Charles Darwin's theory of evolution, I realized further there were many, many more disasters, much, much more serious, than those in recorded human history, at home and abroad. I am convinced that life is indeed a miracle. When I

---

[2] A more detailed research work can be found in his monograph, Ge Jianxiong. 1991. *Zhongguo Renkou Fazhanshi* [History of Chinese Population]. Fuzhou: Fujian Renmin Chubanshe.

read Ralph Waldo Emerson's small poem in *Nature*, I liked it immediately; now I read it with my students every year when I teach a course on American literature:

> A subtle chain of countless rings.
> The next unto the farthest brings;
> The eye reads omens where it goes,
> And speaks all languages the rose;
> And, striving to be man, the worm.
> Mounts through all the spires of form. (Emerson 2009, 18)

Charles Darwin's theory, Ralph Waldo Emerson's words and many other people's good works helped me to realize that not only all people in China but also all human beings are of a family. The heritage chain that leads to me is not only the product of people in China, but also that of the whole humanity. It is true for me as well as for anybody else. Before humanity walked out of Africa to all different continents, they must have lived together and survived many disasters with constant help to each other. Before they transformed into the form of humanity, there had been a much longer history during which they must have suffered even more hardships together. Every time when I meditate on this topic, I realize how much I owe my life to humanity, and it seems I hear a call to urge me to do good to humanity and to treat the world kindly, though I know what I can do is so humble.

In May of 2001, I was invited to read a poem at the Cambridge Conference of Contemporary Poetry (CCCP), and I read *A Love of Shining Night*, a poem that was later on published as a booklet by CCCP. (Li 2002) After reading it, Mr. J. H. Prynne said: "It is a dramatic piece, maybe even apocalyptic."[3] Indeed it is an imaginary piece about the catastrophic consequences of modern science and technology on humanity. I did not know why I wrote and read it then. Now I understand that I was perhaps born like a canary in the coal mine, one who is more sensitive to dangers approaching, and so my mission is probably to urge people to escape these dangers.

In contrast to me who seem to be always worrying too much, my father does not worry about anything at all. He has been living his life in a take-it-for-granted manner, either good or bad, in fortune or misfortune, very much like Rip Van Winkle, a simple good-natured fellow. "If left to himself, he would have whistled life away in perfect contentment." (Irving 1894, 18).

My father neither reads nor thinks much as I do. He received only a few years' school education. His life seemed gloomy before he was recommended by a kind-hearted senior in the village to join the army where he received more education. He wasted no time and learned to write beautiful handwriting. I don't know how he managed to do it, as it seems impossible to me. Occasionally, he would peep at my handwriting that looked terrible to him and sighed loudly behind me. A few times, he proudly offered to teach me how to write beautiful characters, though I firmly declined.

After service in the army of around five years, he was offered a wonderful job in the county government that could have made him a celebrity in the village. Unfortunately,

[3] It is quoted from J. H. Prynne's email to me on April 24, 2001.

he did not actually get the job because a national movement was suddenly launched. Nevertheless, he, respected as a veteran, was given a job in the village management, and later the town government.

After the Reform and Open-Up Policy adopted in China in 1978, my father lost his job in the town government, and was appointed as Head of a small liquor factory of five people, owned by the village. It was a great opportunity for him to make a fortune. However, he had no idea how to run a factory. To make it worse, he became addicted to gambling. I have no idea when he started to gamble. I recall he gambled a few times before becoming Head of the liquor factory. I assume he got fascinated by gambling in his childhood. Poor as he was, gambling must have seemed to be a most appealing way for him to make some money. The evil seed was thus planted. He should not have gambled much before going to the army as he was penniless, and he must not have gambled at all in the army, as his life there was strictly regulated. He might have started to gamble occasionally after he started to work and earned some money, trying his luck as well as for entertainment. Since he had never had much money, it was not a serious problem. However, after becoming the Head of the liquor factory who was granted access to the capital of the factory, the problem became very serious.

He started to gamble a lot! In fact, he seemed to be gambling day and night. It was terrible for the whole family. I started to hear my mother quarrelling frequently with him. I remember I met him, eyes bloodshot, early one morning on my way to school. I was later told by my mother that he had gambled for three days and nights continuously. A few times, my elder brother and I were sent by my mother to look for him to bring him back. We wandered in a few neighboring villages, not knowing where to locate him. As he had money and was an easy target in the eyes of other gamblers, he was very much welcomed by those who knew where to gamble without being interrupted.

After about one year, my father lost his job. An audit carried out following this found him responsible for a lot of debt. I do not know exactly how much debt, since he never said it. I assume it was tens of thousands of yuan. It was huge, as the average annual income of a family those days was just a few hundred yuan. The world seemed to have collapsed for the whole family!

I was in primary school when this tragedy happened. Fortunately, my mother did not quit the family. She continued to feed us, though there was almost no meat on the table for a few years. One day, a kind relative brought us two and half kilo of pork and urged my mother to cook it. However, after the relative left, my mother sent the pork to the market for sale as she was in terrible need of some cash. My mouth was watering, but I did not say a word. Thanks to the Reform and Opening-Up Policy, the life standard of people in my village improved rapidly, except, unfortunately for my family that was now struggling in poverty.

I was always good at math in primary school, though I had never taken my schooling seriously. Neither did my family. Nevertheless, I started to work hard upon my teachers' encouragement in the last year of my primary school. I made rapid progress and excelled in the graduation examinations, and was the only student qualified to enter in the key junior middle school in town.

I started to learn English in the first year of the junior middle school, which was my least favorite subject. Nevertheless, I had already had a dream to become a university student, so I worked hard on all subjects, including English. In the graduation examinations three years later, I was ranked a top student, something like the third to the fifth in excellence.

I could have entered a good senior middle school and then a good university. However, I chose to enter a Normal Teachers' School because it offered an iron bowl, a metaphor of a government-secured life-long job. As soon as I entered the Normal Teachers' School, I was entitled to monthly governmental allowance of twenty-eight yuan that was denied to students at senior middle schools. It was quite enough to feed me. I was short and thin when I graduated from the junior middle school. However, when I graduated from the Normal Teachers' School after three years, I was taller than the average and was as strong as an athlete. In fact, I was in the third and the second places in the 3,000 meters and 1,500 meters races respectively in the last year of the school.

Unfortunately, students at Normal Teachers' Schools were not given opportunities to compete with those at senior middle schools to enter a university. Instead, we were given a job to be a teacher. Therefore, I became a middle school teacher at the age of eighteen. Since there was no English teacher in the school where I started my work, I volunteered to teach English, a career I have stuck to up till now. At the end of the first month of my work, I got my monthly salary of 53.5 yuan, around six to seven American dollars, which was the average income those days.

My family's financial situation had been much improved by then. After my father fell into deep debt and became impoverished, his gambling friends abandoned him, so that he had neither money nor partners with whom to gamble. With the help of my elder brother, who went home after graduating from a senior middle school, my father opened a family-based liquor factory that brought good profit, which enabled him to pay back most of the debt in a few years. One year before I started my work, my elder sister followed her peers to Shenzhen, a fast-developing city neighboring Hong Kong, to work in a clothing factory, where she could make hundreds or even over a thousand yuan a month. In the year when I started to work, the debt of my family had dropped sharply to some hundreds. At the end of the first semester of my work, a debt collector went to our home to collect money before the Spring Festival. Out of embarrassment, I paid him off with my savings when I heard it was only a few hundred yuan. It was the last time such a collector would visit our home before or during the spring festival, when families are supposed to gather together to celebrate the start, as well as hope, of a lunar new year. Since my father was no longer penniless, his gambling friends came back to him and his old habit returned. Nevertheless, because my brother was keeping a close eye on the finance of the family liquor factory, and both my sister and I had got a job, the consequence of my father's gambling was not serious anymore. The only victim left was probably my mother who had to watch her savings very carefully.

When I started to teach English, my own English background was very poor, because English was not taught in the Normal Teachers' School. Therefore, I enrolled as an English major student in a Radio and TV University which offered us courses, sold us textbooks, yet never sent us an English teacher. We had to teach ourselves, and to pass all exams to get an associate degree of English. I worked hard and passed all exams and got my degree in three years.

My dream of entering a university was always there. The Radio and TV University is only a university in theory, which did not satisfy my thirst for becoming a good English student. After getting the associate degree of English, I was qualified to take exams for entering a college via the college entrance examinations for adults. I took it in the fourth year after I began my work, and it was a success! I became an adult university student!

It was a pretty good university where I met teachers of English who were kind and professional. Because I had learnt English with the so-called Grammar-Translation Method, I was able to read English texts with the help of a dictionary, though my listening comprehension was poor, and my spoken English even poorer. What I had learnt was a kind of deaf and dumb English, as it was called. To my joy, I found there were three teachers who were native English speakers, two senior males from the US and one young girl from Belgium, all very kind to their students. I felt like a saying popular in China "a sweet shower after a long drought," and took every opportunity to practice my English with the three teachers from abroad, as well as with other teachers and my classmates.

I made rapid progress and felt very much satisfied. Young and energetic as I was, I never felt tired of learning English. In contrast to the four years' lonely experience of learning English by myself, it was heaven to me to learn English in an actual classroom with classmates around and a teacher lecturing. I started to enjoy learning English. I gradually became fascinated by the ideas expressed in English as well as the people who spoke it. After two years, I got my BA degree. Meanwhile, I took the examinations for becoming a graduate student, and I succeeded! I also fell in love with a charming classmate who very soon became my wife. Later on, I learned her father lost both his parents in his childhood.

I continued to work hard on English. However, I felt I made much slower progress after entering in the graduate program. I did not realize what was happening and just did whatever I could. I now understand the reason I was much less rewarded during my graduate years: it was because of the English educational methodology, i.e., the so-called Grammar-Translation Method, adopted by my university as well as most universities in China. Nevertheless, after three years, I graduated with a MA degree.

I got a job in Guangzhou University, where I have been working up till now. The first class I taught was of first-year students with whom I got along very well. After a few months, I was asked to teach a course to a group of senior students. It was a disaster because I found quite a few students spoke much more fluent English than me, and a good portion of the class seemed not to enjoy my lectures. One student openly challenged me in class, and a few mocked me among themselves in a kind of whispering that was loud enough to be heard by me. Once again, I felt embarrassed and ashamed of myself. I thus decided to turn myself back into a student for another breakthrough of English-learning. After one year, I became a PhD student in Sun Yat-sen University and met my PhD Director, Prof. Ou Hong, who was to help me to transform into a scholar of poetry in English.

With help of Prof. Ou, I was granted a scholarship from CSC (China Scholarship Council) and, with his recommendation, I was accepted by Mr. J. H. Prynne as a visiting scholar at the University of Cambridge for one year. In the meanwhile, Prof. Nie Zhenzhao wrote a warm recommendation letter on my behalf to Dr. John Rathmell, another remarkable senior scholar at Cambridge. Mr. Prynne took the responsibility to supervise my doctorate thesis, and he offered to meet me at four o'clock on every Thursday, and often discussed my work with me until dinner time. He sometimes invited me to have dinner with him and continued to talk with me until deep in the night. The next day, when I returned to his upstairs study, there was always a pile of books waiting for me, sometimes together with some pages of my thesis corrected in red ink if I had submitted them to him the day before. Dr. Rathmell offered to read modern poetry in English with me for an hour at noon every Monday, which helped me to truly understand the beauty of modern poetry.

I was so lucky to have met Mr. Prynne and Dr. Rathmell, who, as I believe, offered me the best education available in the world. It was with their help that I became completely confident when I came back to the classroom as a teacher of English. Paradoxically, it was by working on ideas, as well as literary beauty, carried in the English language rather than on the boring grammatical rules, the core of the so-called Grammar-Translation Method, that I truly mastered the language. It is the most important lesson I learned, which would afterwards lay the very foundation of my philosophy of English education.

I often felt a little guilty as I could not afford to pay either Mr. Prynne or Dr. Rathmell for their marvelous work. I was so short of money that I sometimes took just some biscuits for my lunch. To save money, I would only buy myself half a pint of beer when I went to a bar with my friends, some of whom were kindly introduced to me by Mr. Prynne. I felt grateful that neither Mr. Prynne nor Dr. Rathmell asked to be paid. In fact, they never mentioned a word of it! If they had, I would have lost the courage to ask for supervision from them. After some time, I realized their generosity was out of the true spirit of a scholar as well as of poetry, a spirit that I have taken up as my academic and poetic faith. I tell the stories of their kind help to me to illustrate the true spirit of a scholar and of poetry every year when I teach the course of *Selected Readings of Modern Poetry in English*, which has now been made in to a MOOC (massive open online course).[4] I also published an essay that reflects on this experience and discusses the philosophy of education I mentioned above, in which I argue that the best way of education is dialogue (Li 2015).

I had already distanced myself from the stereotyped image of all westerners as monsters, an image that was repeatedly presented in the stories of western powers' invasion of China in the modern age, after I had read some books in English, especially after my personal experience with these three teachers from abroad. After I got to know Mr. Prynne and Dr. Rathmell, and after I read more books to have a much more profound understanding of the western culture as well as the happiness and sorrows of the ordinary people in the west, the stereotyped image of western people was totally dissolved, as I found people in the west are really the same as those in China.

[4] The website link is: https://www.ulearning.cn/course/25598.

Furthermore, after knowing how important as well as how difficult it is to overcome a stereotype, I became vigilant of all sorts of stereotypes so as not to be fooled by them. I had found some of my fellow people were suffering from some stereotypes without being granted good opportunities to overcome them. In due course, I found some people in the west were also victims of certain stereotyped images of either China or its people. I would always give a hand to help whenever there was an opportunity. In fact, I have found that many problems in the present world, such as those of racism, nationalism, sexism and so on, are caused by all sorts of stereotyped ways of seeing the world. Racists who are rampant in certain areas in the world probably suffer the most from stereotypes that make them blind and keep them living in darkness.

Besides meeting with Mr. Prynne and Dr. Rathmell, I took as many lectures and seminars as possible, trying to learn as much as possible, and took every opportunity to interact with all sorts of people I met, though, to be frank, I was sometimes at a loss as to where I was. I made great progress and would identify myself as a scholar, though not very advanced, when I left Cambridge for home in early 2002.

In 2016, I was granted another scholarship by CSC to visit The Kelly's House at University of Pennsylvania as a visiting Professor. With kind help of my host Prof. Charles Bernstein, I published a booklet of poetry with the title of *Zhongalish: Feel and Think Globally*. (Li 2016) I coined the word *Zhongalish*, i.e., *Zhongguoren*, American and English, to promote the idea that one must overcome the stereotype of nationalism to learn from all cultural heritages created by all people in human history. Indeed, all founders of cultural values and practices are keen to welcome people, despite of their nationalities, to learn from them. In a certain sense, we can say that all human cultural heritages first and foremost belong to humanity rather than their respective founders or the nations they lived in, though the founders should be thanked and honored for their work. Only by learning from all human cultural heritages, can one take full advantage of the modern age and get oneself fully cultivated, becoming a qualified modern person so that one could better serve modern humanity.

Thanks to the Reform and Opening-Up Policy and all people at home and abroad who made it a great success, my salary had been raised a few hundreds of times since I started to work, so that I was able to bring my son with me for this visit to the US. In order to help him to improve his English so that he was able to well communicate with people around and would not feel isolated in class, I worked hard to sort out of an efficient method. Eventually, inspired by my experience at Cambridge, I decided he should take the method of extensive reading of English literary works. Therefore, I invited a wonderful lady, Ms. Katherine Ernst, who had just graduated from Swarthmore College, to read novels in English with my son. It turned out to be a tremendous success! Within three or four months, they completed close reading of all the seven books of *Harry Potter*, and my son became completely confident in his classes. We did not stop there. I asked my son to continue to read literary works in English until he graduated from senior middle school, with help of Ms. Katherine Ernst first and Prof. Neil Roberts later. By then, he had read more than sixty books in English and grew to be a very fluent English speaker.

Upon the request of some friends who asked me to help their children with their English, I initiated a program called MEEP (Modern English Education Program) which was very successful. All students who worked hard in this program turned out to be confident English speakers. All of them went to good colleges; some were offered a place by top universities at home and abroad. I was overjoyed to have found an efficient way to help students in China to truly master English, and published an essay to share this joyful method with my colleagues. (Li 2024).

My son received offers from a few top universities at home and abroad when he graduated from senior middle school. He wanted to choose to go to the University of Cambridge for college education. My wife and I discussed and decided to back him up. Though it meant we had to spend up all our savings, we considered it worthwhile because we believed Cambridge was the best possible place for him to fully develop his potential. Besides, since our income is paid by the world that is connected by "a subtle chain of countless rings," it is probably the best way for us to pay it back to the world for education of the next generation, so that they could provide the best service to the world in the future. I am an educator all my life, and I do believe that education is the most important in the world which deserves as much investment as we can possibly afford to pay.

In due course, I realized that the key for modern education, including English education, is the inculcation of modern cultural values, so that I elevated MEEP to MECEP (Modern English and Culture Education Program). In the meantime, I listed and explicated some key modern cultural values for personal cultivation for my students. As there are sixteen weeks each semester in China, I selected sixteen of them, i.e., (1) to love life, (2) to be kind, (3) to be honest, (4) to be organized, (5) to be self-disciplined, (6) to be independent, (7) to be open-minded, (8) to be brave, (9) to be grateful (10) faith in reason, (11) faith in knowledge, (12) faith in equality, (13) faith in fairness, (14) faith in humility, (15) faith in work, (16) to be resilient. Of course, there are many more key modern cultural values in personal cultivation and other perspectives, such as freedom, democracy, rule of law and so on that have been well explicated and promoted. I do not intend to give an exhaustive list of key modern cultural values. Instead, I just want to work on some of them to help my students to get some initial sense of modern culture: that is, to cultivate in them a mindset of modern culture. When they grow up, they will probably figure out their own vision of modern culture and their own definitions of key modern cultural values that might be even more sophisticated than mine. Whatever their definitions are, they will be serving the modern world in their unique ways, so long they are good at thinking in the mindset of modern culture.

Modern cultural values are good not only for guiding my students and myself in our lives, but also for curing myself of certain problems that had troubled me quite a lot. For example, I had realized that I sometimes felt unconfident when I should have been confident, and that I sometimes felt a little anxious when I shouldn't have, and so on. After learning of Trauma Theory in literary studies, I came to realize the problems were probably caused by my personal experience in my childhood that was much troubled by my father's gambling. It is said such problems are extremely difficult to cure because they are deeply rooted in our life experiences, and they cause the worst consequences if we do not realize their existence or do not guard against them. I also came to realize that my father's excessive addiction to gambling was probably caused by his sufferings in his childhood, which made it much easier for me to forgive him. In fact, I have been sending him a little more money than necessary all these years so that he could play a little bit of mahjong, a gambling game, with his companions. His gambling habit brings no more harm to anyone now and it brings him a lot of enjoyment. It could probably help protect him from Alzheimer's disease too.

Trauma can be passed from one generation to the next if it is not identified or effectively treated, as is surely the case with my father passing his to us. It was most unfortunate that my father never realized this problem: like many other people in the world who unwittingly bring their sufferings to the world, their family members often being the easiest targets.

I noticed that my PhD supervisor wrote and talked a lot about hardships he suffered in his childhood as late as in his sixties. Later on, I realized it served as a good means for him to cure himself of his childhood trauma, which, according to my observation, was very successful. Fortunately, I had realized my own problems and have found the best way for me to cure myself of my childhood trauma is to get myself cultivated with key modern cultural values. It helps us to overwrite all sorts of mechanisms that were formed up in our life experience, including those in my childhood. Such mechanisms guide us to respond to the outside world in a kind of seemingly intuitive way, which I often remind myself to take another thought in the light of modern cultural values, especially when it is something of importance. In this way, I often adjust myself when my intuitive response is not quite right, so that I can enjoy my life much more as modern cultural values are themselves designed to promise their followers a happy life.

My elder sister has been working in Shenzhen all her life. My brother joined her after running the family liquor factory for a couple of years. Although they have been making ten times more than whatever they could make at home as farmers, their income is low in comparison to the cost of their livelihood and their family responsibilities. As they are farmer workers, they had to leave their children at home, taken care by their grandparents, which was the only affordable way for them. They have been working hard and are saving, but it seems they are always short of money. To give them a hand in a discreet way, I have been covering the cost of my parents' livelihood and medical care all these years, so that they don't need to pay them. I do not know how my elder brother and elder sister dealt with their childhood trauma, or whether they have ever realized its existence. Since they did not receive higher education, they probably have never thought of it in the way I did. It seems not quite feasible for them to apply the delicate way of self-cultivation with modern cultural values either. They might have to find their own ways to deal with it, or live with it all their lifetime.

Both my wife and I love our son enormously and have been providing him the best in all aspects as we could afford. Since both of us are college teachers, we pay a lot of attention to the education of our son who, both polite and bright, has done much better than we had expected. We would very much like to always keep him close to us, but we understand he is an independent human being and is entitled to live a life of his own freewill, cultivated with modern cultural values, so that we have supported his decision to go to take a college education abroad.

My son and his generation are much luckier than the former generations, though they also need to work hard to find their own ways to a happy life. Besides wishing them the best, what I could do to assist them is probably to offer them some advice from a scholarly perspective in my field. Indeed, while I am writing this book, my son and his generation, as well as their friends at home and abroad who would like to know more of modern China, are taken as a group of the most important target readers. Hopefully, this book can help them to understand how sad China once was, how hard it has been for China to transform from the traditional to the modern, and how important it is to help ourselves and whoever we care to be well-informed of and well-cultivated with modern cultural values, and to keep on moving forwards on the right track! Of course, I understand my son and his generation are all independent-minded, so that I won't feel extremely disappointed if they don't agree with me at some points. I am happy to have done what I take as my duty as a member of the last generation as well as an English teacher, a career that was more assigned to me by Heaven than chosen by myself.

Last but not least, I would like to take this opportunity to thank Dr. Daniel Braun for accepting my invitation to undertake this project with me, and assisting me throughout the writing of this book. Our cooperation is prefect!

Zhimin Li, August 29, 2024.

Works Cited

Emerson, R. W. 2009. *The Essential Writings of Ralph Waldo Emerson.* Brooks Atkinson, ed. New York: The Modern Library.

Ge, Jianxiong. 2008. Zhongguo Lidai Renkou Shuliangde Yanbianji Zengjiande Yuanyin [The Changes of Population and Their Causes in Different Dynasties in China]. *Dangde Wenxian* [Literature of Chinese Communist Party] 2: 94–95.

Li, Zhimin. 2015. 'Duihua' Jianqiao: Youguan 'Duihua'de Gushiyu Lilun [Dialogues at Cambridge: Stories and Theory of Dialogues]. *Xueshu Pinglun* [Academic Review] 5: 80–85.

Li Zhimin. 2002. *A Love of Shining Night*, Kevin Nolan. ed. Cambridge: Cambridge Conference for Contemporary Poetry.

Irving, Washington. 1894. *Rip Van Winkle.* Boston: Joseph Knight Company.

Li, Zhimin. 2016. *Zhongalish: Think and Feel Globally.* Philadelphia: CAAP.

Li, Zhimin. 2024. Yingyude Lingyiban: Yi 'Yingyu Wenhuatixi' Weizhongxinde Yingyujiaoyu Lilunyu Shijian [The Other Half of English: Theory and Practice of English Education Centered on the English Cultural System, Foreign Language and Literature]. *Waiguo Yuwen* [foreign Languages and Literature] 2: 174–185.

Lu, Yang. 2019. Qishinian Zhongguoren Pingjun Shoumingde Bianqian [The Changes of Lifespan of People in China in the Past Seventy Years]. *Xinmin Evening News*, December 4.

Chapter 2
The One-Way Model of Cultural Interaction: Literary Interactions Between China and Cambridge

An examination of what happened in the past can often shed light on what is happening and what will happen. This is demonstrated by studying the interactions between Chinese and English literature, with Cambridge as a famous base, disclosing some unique characteristics of what may be called the one-way model of influence, which can be explained only when the historical background is taken fully into consideration.

A Brief Review of Interactions Between China and the West

China and the West had developed into two distinctively different civilizations independent of each other before any recorded interactions occurred. There is not a single verse acknowledging the existence of China in the *Bible*, the founding text of Western Judeo-Christian civilization. It is generally believed that it was Marco Polo (1254–1324) who, by telling many wonderful stories of China, popularized its image in the West—though disputes exist as to whether he had actually been to China, as he claimed. By the seventeenth and eighteenth centuries, China had come to be regarded by European nobility and members of literary society not only as a venerable civilization in its own right, but as a model to imitate. The French philosopher Voltaire wrote: "it is true that the constitution of their empire is the best in the world, the only one entirely based on paternal power." (Voltaire 1972, pp. 114–115).

Indeed, European elites were often keen sinologists, furnishing their studies after the Chinese style, or dressing in the Chinese fashion. It was recorded that many attending a masquerade ball hosted by the French court on the first New Year's Eve of the eighteenth century had outfitted themselves in Chinese fashions in a display of their refined taste (Hao et al. 1996, p. 280). Chinese customs, such as tea-drinking, gradually took root in Western culture as well.

Z. Li and D. Braun, *China Being Led and Leading*,
https://doi.org/10.1007/978-981-96-6803-8_2

In such an atmosphere, Europeans learned quite willingly from China and achieved some successes. As Jacques Gernet put it:

> China furnished the first example of a disciplined, rich, and powerful state which owed nothing to Christianity and seemed to be based on reason and natural law. It thus made a powerful contribution to the formation of modern political thought, and even some of its basic institutions were imitated by Europe. The Chinese "system of examinations" was described for the first time by Mendoza in 1585 ... The same institution was to be extended to Great Britain in 1855 with the adoption of examination to the Civil Service. (Gernet 1982, p. 523)

Joseph Needham, another well-known Sinologist, brings us much insight within an even wider field of cultural interaction between China and the West. Agreeing with the philosopher Francis Bacon that printing, gunpowder, and the magnet had "changed the whole face and state of things throughout the world," Needham concluded that all of these technologies "were due to Chinese experimentalists." (Needham 1986, p. 9) Yet China meanwhile showed little knowledge of or interest in the Western world, having long regarded itself as the most powerful and civilized nation on earth. During the sixteenth century, missionaries from the West, led by Matteo Ricci, began to win the respect and friendship of Chinese contacts, successfully converting some portions of the Chinese population to the Catholic faith, while introducing them to central texts of the Western tradition. Indeed, during the sixteenth and seventeenth centuries, over 150 texts from Greek and Roman antiquity and early modern Europe were translated into Chinese, among them *Euclid's Elements*, translated by Ricci himself, in collaboration with the Catholic Chinese scholar Xu Guangqi (1562–1633). These translations are the most significant achievement on the part of China during this phase of cultural exchange with the West.

As more missionaries entered China to convert its population, significant disputes over the compatibility of the Catholic faith and Confucian practice arose. Popes, from Benedict XIV to Pius VI, insisted that Chinese Catholics give up the worship of ancestors, a key moral value in Chinese culture that cut across religious lines. Emperor Yongzheng (1678–1735), supported by Confucian Chinese scholars, insisted that ancestor-worship was an indispensable value in Chinese culture, which was to be universally observed. The clash caused Emperor Yongzheng (1678–1735) to issue orders in 1734 prohibiting missionaries from converting any Chinese to Catholicism. The result of this proscription of Catholicism was the gradual withdrawal of many missionaries from China, as well as a rapid decline in the Chinese assimilation of Western culture.

The arrival of Ricci and his fellow missionaries and Emperor Yongzheng's crackdown on their activity mark the boundaries of the first phase of the cultural interaction between China and the West. The period can be characterized by three distinctive theses: (1) both sides considered themselves equal to the other; (2) the basic cultural beliefs of both sides remained intact while they learned from each other; (3) both sides took the position of learning from the other with the aim of serving their already existing ways of life. In general, the first phase witnessed a set of free and equal cultural interactions, though there were drawbacks: for example, China had been largely ignorant of Western civilization prior to this first contact, and the channel of communication between the two civilizations was artificially narrow, and asymmetrical, with Catholic missionaries constituting the chief conduit between China and the West.

The Opium War of 1840, waged by Britain against China, brought an unprecedented change in the relationship between China and the West. Indeed, scholars now generally regard 1840 as marking the birth of Chinese modernity. In fact, the Opium War set off a domino effect in China leading to a dramatic change of civilizational course:

> The "self-strengthening" movement, as it was styled, proved a dismal failure; already discredited by 1871 it was finally and totally wrecked in the disastrous Sino-Japanese War of 1894. China's faith in herself thereupon collapsed entirely ... The Chinese began to doubt anything of their traditional culture and this gradually caused the May-fourth movement that was to overthrow all traditional Chinese cultures and rebuild their culture by learning everything from the West. (Dawson 1964, pp. 80–81)

This is to condense a very complex story; but the phrase "learning everything from the West" is emblematic of the second phase of the cultural interaction between the two great civilizations.

The second phase, kicked off by the Opium War, is characterized by significantly different dynamics than those of the first phase. First, the mutual respect and equality among the Chinese and Western emissaries gave way to an imbalance in which Western supremacy produced a reciprocal Chinese inferiority. The Japanese scholar Mitsuishi Zenkichi has observed that while Europeans admired China throughout the seventeenth century, "more and more Europeans started to show contempt towards China around the eighteenth and nineteenth centuries." (Zenkichi 1998, p. 9) Britain's victory over China in the Opium War, along with a cascade of subsequent western military victories, gave this 'contempt' a distinct color and foundation. Second, while China as a nation faced an existential challenge, many basic Chinese cultural values were questioned and given up in panic by leading Chinese scholars, who called on the general public to do the same. Meanwhile, the leading Chinese scholars introduced all sorts of Western ways of life to replace Chinese traditions, as they believed it was the only way to help China to survive and to revive. The situation was recorded in the Chinese language itself, with the adjective "old" (旧—often connected with the traditional) taking on the connotation of "bad," and "new" (新—often connected with the Western) "good." These shades of meaning would prove ineradicable, persisting through the twentieth century and into the present day. Needless to say, no corresponding shift of this extent, driven by Chinese influence, occurred in the west. Third, the aim of "learning everything from the West" was to save China as a nation, which caused the abandonment of traditional Chinese culture; this, in turn, resulted in China losing any solid cultural criteria for choosing what, and what not, to learn from the West. This caused great cultural confusion that inevitably led to many cultural upheavals in China in the twentieth century, as represented by the May Fourth Movement of 1919. The effects caused by these many upheavals are felt right up to the present.

It is clear that both China and the West were independent in the first phase of contact, and their cultural interactions generally followed a two-way model of exchange. However, in the second phase, China lost its cultural independence under the pressure and influence of the West, especially in the military and political arenas, leading it to adopt the practice of learning from the West whatever it could to replace, rather than support or supplement, its own cultural traditions. And it was during this phase that the two-way model of cultural interaction gave way to an asymmetrical one of cultural interaction that can justifiably be called a one-way model.

The One-Way Model of Literary Interaction

We are social beings, circumscribed by cultural boundaries, defined by different perspectives at different levels: ethnic, linguistic, ethical, and so on. At present, nationality is among the categories we most often and easily reach to for social definition. The construction of a certain cultural boundary can benefit those within it in many ways, but cultural boundaries are never static. On the contrary, they are in constant flux, sometimes gradual, at other times, dramatic. This cultural dynamism is common to the West and China. The general stability of a culture, often during stages of very gradual change, is ensured by two preconditions: (1) the balance of power of different interest groups within the culture, and (2) the ability to resist aggression and influence from without. At certain moments, struggles between different interest groups can result in a splitting of one culture that fragments into several distinct and even rival cultural identities. On the other hand, the combination of different cultures, by peaceful or violent means, can produce one larger culture.

The interaction of cultures often transpires at two levels: the level of ideas, and the level of force. The first phase of cultural interaction between China and the West occurred mainly in the realm of ideas, while the second phase witnessed the active involvement of much military force. Literary interactions, which might seem to be independent of military and political struggle, in fact are not. Indeed, the two phases of cultural interaction between China and the West are a case in point of the entanglement of these seemingly autonomous levels of exchange.

China is a nation with a long and rich literary history. Its pride and insularity (along with its linguistic isolation) thwarted interaction with Western literature during the first phase of cultural contact. The books translated into Chinese during this phase belonged primarily to the fields of natural science and religion. Not so on the Western side, for a significant number of Chinese literary works were translated into European languages and were greatly appreciated: Voltaire praised *The Orphan of China* (《赵氏孤儿》), and Johann Wolfgang von Goethe claimed *The Fortunate Union* (《好逑传》) a masterpiece which contributed to the formation of his concept of world literature.

The involvement of military and political powers dramatically shifted the dynamic of exchange in the second phase. Literary interactions often begin with translation. Yet very few literary works were translated into Chinese, even at the beginning of the second phase, as most Chinese in their effort to preserve their national autonomy, believed that what they needed to do to strengthen China was to learn from the West natural sciences, especially military science. Hu Shi, one of the leading figures of several twentieth century cultural movements in China, wrote that "the other books such as literary books and philosophical books were ignored in those days. It was natural for Chinese then to think that Western art and philosophy would certainly be lagging behind China, with its 5000 years of history, though the Westerners had very powerful guns." (Hu 1999, p. 93) It is hardly surprising that the self-confidence of Chinese scholars in their literary tradition should have allowed it to persist unmixed with Western influence for so long.

The selection by Chinese scholars of the Western texts to be translated was dramatically altered by the failure of the Sino-Japanese War of 1894–1895. China's defeat convinced scholars and advisors that merely strengthening the nation's military forces could not save the country from imperial conquest. The shift was abrupt, with leading intellectuals calling for the nation now to begin to learn from the West in an all-inclusive manner. Few Chinese scholars, however, were competent to translate works from their original language—a by-product of China's self-isolation. As a workaround, they were able to introduce works of Western philosophy, politics, economy, literature, and more by translating those for which Japanese editions existed. "During the three hundred years before the Sino-Japanese War of 1894–1895, the Chinese translated only 12 Japanese books. But within the 15 years after it, the Chinese translated 958 Japanese books, most of which were originally Western or written about the West." (Zheng 1999, p. 312) The routing through Japanese can be said to mark the formal beginning of the process of the modernization of Chinese culture, a process that is not yet finished.

In the field of literature, it was Hu Shi, who started and led the literary revolution in China at the beginning of the twentieth century, openly proposed imitating the West, going so far as to announce: "after very careful thinking, I have got one idea, that is: to translate as much foreign literature as possible to serve as our model." (Hu 1999, p. 47) Hu Shi himself led the way with the publication of revolutionary and western modernist inspired New Chinese Poetry. Hu Shi chose poetry for a reason: the verbal art had been regarded as the most noble of all traditional Chinese literary genres. A change in the field of Chinese poetry would thus rock the very basis of Chinese literature as a whole. Hu Shi achieved great success, yet his efforts to overturn tradition would have been unthinkable without parallel victories by Western powers over China in other fields. (Li 2005, pp. 13–22).

The history of New Chinese Poetry reveals at every turn its Western characteristics. Hu Shi claimed that his poem "Cannot Shut In" marked the maturation of New Chinese Poetry. The poem is in fact a Chinese version of the American poet Sara Teasdale's "Over the Roofs." This act of poetic translation and transposition is an elegant symbol for New Chinese Poetry. (Zhang 1998, p. 33) With Hu Shi as their example, many poets sought models in Western poetry, leading eventually to its general admiration. Extremists such as Ji Xian (1913–2013) went so far as to propose a so-called "horizontal transplantation" of Western poetics into China, although this proved unsuccessful, as he later confessed.

In the twentieth century, Western poets also learned from Chinese poetry; interestingly enough, it was the very traditions of Chinese poetry that had been "discarded" in their native land that were most successfully implemented by Western poets. The most famous example of western translation and transposition of outmoded Chinese poetry is that of Ezra Pound, who initiated Imagism and refined the genre with his publication of *Cathay* (1915), a work which collected numerous Tang era poems, rendered in his signature economical style. The key to Pound's success lay in his ability to learn from Chinese poetics and adapt it to Western aesthetic concerns. His aim was to advance his native medium and tradition: this differed significantly from Hu Shi and his followers, who introduced Western poetry in order to supplant traditional Chinese poetry, a project more destructive than constructive.

In the second phase, China suffered an unfortunate cultural collapse that caused an uncontrolled influx of Western literature, resulting in the prevalent one-way model of literary interaction. In that particular historical circumstance, there was no independence in all fields for China, which had no way of assuming the role of an equal partner to the West.

Xu Zhimo and William Empson Crossing the Boundaries

The characteristics of literary interaction between China and the West in the second phase can be vividly demonstrated by two cases, namely, those of Xu Zhimo (1897–1931), a Chinese poet who lived in Cambridge, and William Empson (1906–84), a Cambridge poet who lived and taught in China from 1937–38, and again from 1947–53. The difference between the status of Xu as a student in Cambridge, and Empson as a teacher in China, is crucial to their respective roles in the one-way model of literary interaction.

Originally, Xu was encouraged by his family to study economics in the US, in order to help "save" China, a mission not uncommonly promoted among his social class. Yet Xu simply did not feel any passion or commitment to the field. He once said of himself that "if he had been a person with nothing to be said for him when he arrived in the United States, nothing in him had changed when he left America." (Zhao 1989, p. 19) In 1920 Xu left the US for England and became a visiting student at King's College, Cambridge. It was here that Xu's poetic talent was aroused and developed, a talent that would lead him to become one of the most admired New Chinese Poets of the twentieth century. Xu himself said: "I was indeed blessed in Cambridge, a sweet life experience that, I am afraid, would never come back again … my self-consciousness, was nurtured by Cambridge, my appetite for knowledge was awoken by Cambridge and my eyes were opened by Cambridge." (Xu 1927, p. 42) Cambridge became Xu's spiritual home and a central image in his poetry and poetics.

Xu's most famous poem is "Farewell Again to Cambridge," a poem often recited, and almost certainly familiar to the vast majority of educated Chinese people. Indeed, it played a significant role in shaping the Chinese image of Cambridge as a modern-day Eden.

Quietly I am leaving as
Quietly I have come
Quietly I wave farewell
To the gold clouds in Heaven.

The gold willows on Cam River
Are like brides in dusk
And their reflections in the Cam
Wave gently in my heart.

The green plants in the Cam
Are dancing elegantly in the water
In the gentle caress of the Cam
I would like to be turned into a grass

That small pond in a green shade
Is not a spring, it is a heavenly rainbow
That has been melted among grasses
With a sweet and colourful dream.

A dream? To take a long punt-pole
For searching more green in greenness
To carry a boatful of star-light
Singing in the marvelous starry night.

But I could not sing songs,
Quiet is the lute of leaving;
Even the summer insects keep quiet
Even Cambridge is quiet tonight.

Gently I will leave as
Gently I have come
Gently I wave my hands
Not to spoil the light of a cloud. (Xu 1990, p. 317)[1]

In fact, it has been said that "Farewell Again to Cambridge" single handedly convinced the scholar Lin Yutang (林语堂, 1895–1976) of the value of New Chinese Poetry:

> New Chinese novels and drama had soon been accepted by general Chinese readers after the May Fourth Movement; however, New Chinese Poetry still has to be tempered in the future to achieve the same honour. In the past, many scholars who had been accustomed to reading classical Chinese poetry or even Western poetry without writing poems themselves, such as Lin Yutang, did not consider New Chinese Poetry as poetry at all until they read Xu's poems. Therefore, Xu did contribute to New Chinese Poetry by making it live in the hearts of general Chinese readers. (Xu 1983, p. 4)

---

[1] This version of translation is done by Zhimin Li.

Xu soon built a solid reputation as a talented poet widely read in China. However, to an English reader "Farewell Again to Cambridge" may not immediately appear to be a great poem. It is therefore worth examining the work a bit more closely. The vital difference between a talented poet and an ordinary poet is that the former can often create a kind of linguistic music, almost by intuition, that the latter, despite however much hard work, can never match. "Farewell Again to Cambridge's" music triumphs over many other New Chinese Poems, but in order to grasp its artistry, one must note certain features of the original Chinese that don't make their way into the English version printed above.

In this poem, Xu famously adapts the rhyming ABAB pattern of the English ballad stanza, a pattern never before deployed in Chinese poetry. Xu also makes use of another feature of western poetics that had been entirely foreign to Chinese poetry: namely, enjambment, a device that allows him to create more syntactical possibilities among and between lines than would not have been possible with traditional Chinese poems that are not determined by lineation.[2] Yet neither of these two features can be used to distinguish this poem from others that are apparently similar. What makes this poem unique in the history of New Chinese Poetry is Xu's accomplishment in Pianwen, a traditional Chinese poetic writing style that emphasizes the musical quality of a text.[3]

In terms of meter, Pianwen consists of sentences of two or three metrical feet, occurring interactively. In "Farewell Again to Cambridge," almost all the lines can be read as having two or three feet, though not so strictly as in Pianwen, such as the first stanza: 轻轻的/我走了/, 正如我/轻轻的/来/; 我轻轻的/招手/, 作别/西天的/云彩/ There are quite a few lines in the poem that consist of sentences that accord exactly with the requirements of Pianwen. Xu's careful choice of characters grants the poem a sort of natural linguistic melody that is itself expressive, articulating almost the same feelings, through a rich sound symbolism, that are suggested by the images. Thus the meter and images in this poem create a perfect unification of poetic beauty. In fact, there are quite a few famous New Chinese Poems whose success depends on their adaptation of features of traditional Chinese literature, which indicates the problematic nature of the overwhelming one-way model adopted by new Chinese literature. This is one example of the fact that, with the unprecedented Westernization of Chinese culture in all fields, many excellent traditional Chinese cultural ideas and practices have been ignored or driven underground, which we will discuss further in the book.

The popularity of Xu's "Farewell Again to Cambridge" reflects, in its melancholy, the state of mind of the general public in China during the second phase. Admiration for the West came not so much from profound understanding of its culture, or contact with westerners themselves, but from a sense of national humiliation in defeat. Xu's

[2] It is worth noting that traditional Chinese poems are often divided into lines in the modern period, following the practice of New Chinese Poetry, which makes many scholars in the West believe that traditional Chinese poems were also lineated.

[3] The English translation of "Pianwen" is "parallel prose," though this term does not exactly capture the technique's essence.

poem functioned as a kind of wonderful placebo, a work of art that sublimated public sentiment.

A biographical reading of the poem reveals another layer of its mystique. Indeed, Xu's poem is not so much dedicated to Cambridge as it is to Lin Huiyin (林徽因, 1904–55), the woman he met and fell in love with while studying abroad there. The poet's love has become something of a legend in China. Xu, married at the time, decided to get divorced shortly after falling in love with Lin. In August 1922 he left Cambridge to follow Lin back to China. It is perhaps a pity that Lin made a rational choice and married another man in the spring of 1928, which brought an end to Xu's dogged (and public) pursuit of her, though not to his secret love. In November 1928 Xu composed "Farewell Again to Cambridge," in which he enciphered his longing for Lin as a longing for Cambridge. The images of gold willows and brides are in fact the perfect blending of the symbols of Cambridge and Lin in the poet's heart.

William Empson led a completely different life in China from that of Xu in Cambridge. Unlike Xu, who left to study in the West in order to serve his nation, Empson went to China mostly for a job opportunity. "He [Empson] had gone to China not with an especially virtuous purpose but simply to fill a paid post, whereas those educators who set themselves up as cultural missionaries – he had Dorothea Richards in mind as one of a number who spoke in terms of 'We Who Love China' – had withdrawn from the field of action." (Haffenden 2006a, p. 503) Paradoxically, Empson achieved much more success in the field of cultural exchange than did the "cultural missionaries," though he did not mean to. Unlike Xu, who claimed a close emotional tie with Cambridge, Empson did not show much admiration for China. In a poem titled "China," Empson expressed a mixture of sympathy and apathy with respect to the country, a recurrent theme in his comments and memories of his time abroad—though he did speak favorably of numerous Chinese individuals (Empson 2000, p. 90). Again, unlike Xu, who was keen to absorb Western cultural values that could replace Chinese ones, Empson was in general confident in his own cultural heritage, though he was not uncritical of it. In fact, Empson's Western cultural values can be seen to guide his observations on China.

Empson stayed from 1937 to 1939, and again from 1947 to 1952, during which time he had the perfect opportunity to learn Chinese; as a talented scholar, he could have mastered the language had he wished to. Yet he let himself fail after a few fruitless attempts, implying a lack of serious interest in Chinese culture, despite the English books concerning China that he read, and the interactions of all sorts that he had. Empson wrote:

> I see few people, talk no Chinese, and take very little interest in the local arts ... It is rather hard to say why I like being here so much, as I certainly do... No doubt you could generalize it into saying you like to be in a place where there is no sense of sin and no feeling that it is a duty to worry, combined with an adequate amount of hardheadedness. (Haffenden 2006, p. 105)

It can be concluded that Empson did not go to China because of his love of its culture. Rather, he chose a job abroad because he had suffered deep humiliation at home at Cambridge, which left a wound in his heart that seems never to have fully healed.

Despite his apparent lack of interest in the country, its language, and culture, as a professor of English, Empson was deeply admired by his Chinese students. One of them, Wang Zuoliang, who regarded Empson as a role model, recalled with great admiration that the expatriate professor once typed out the whole text of Shakespeare's *Othello* from memory in order to teach it. (Deng 2006) Many other students of Empson's expressed similar admiration. It is worth bearing in mind that quite a few of them, such as Wang Zuoliang, Xu Guozhang, Zhou Yuliang, Li Funing, Xu Yuanchong, and Yang Zhouhan, were leading members in the field of English education and English literature in China, while many others, such as Mu Dan, Du Yunxie, Zheng Ming, Zhao Ruihong, Yang Yi, and Yuan Kejia, were well known contributors to New Chinese Poetry, through whose work the voice of Empson was broadcast powerfully to the general public. At the same time, Empson was also well respected by his colleagues, who were among the best minds in China in those days. John Haffenden records: "Empson would not have agreed with Hough's estimate that he lacked the 'necessary companions' while in China; he believed his sharp, sage colleagues in the refugee university were up to the intellectual demands of any conversation." (Haffenden 2006, p. 505) In China today, Empson is still revered for the impact his teaching made on a generation of writers and thinkers; he is often placed alongside Matteo Ricci in the conversation about the cultural influence of the West in China.

The status of Xu as a student abroad, and Empson as a teacher in exile, serves as a perfect metaphor for the second phase of cultural interaction between China and the West. Xu went to Cambridge to study, Empson came to China to teach.

Conclusion: Balance in Becoming

All human beings belong to the kingdom of humanity, which yet occupies a tiny spot in the universe. It is no accident that people all around the globe share, both physically and psychologically, so many features. Archaeological evidence suggests that nothing like cultural interaction occurred until late in the development of the species. Yet culture, if not in the exact sense we take it now, has always existed for us. And the fact that human beings share a more profound sameness (from our genome to the spectrum of epigenetic features) makes it possible for different cultures to interact with each other. In a certain sense, following the eighteenth century concept of world literature, we can take the totality of all cultures and their languages as the most general boundary of the human species which yet belong together in their variety.

Nothing can stop the interaction between different cultures so long as emissaries continue to bear their heritage forward. The process must be considered from both sides as well. Interactions with another culture can often help and promote the growth of one's own culture. They can also lead to cultural cooperation in efforts to tackle problems locally and globally that were heretofore insoluble. A more inclusive cultural boundary at a different level that contains strong elements of the constituent cultures can sometimes result, which can promote more disciplined competition and smoother cooperation between different interest groups and individuals. These are the desirable outcomes that healthy cultural interactions, which are often part of a two-way model, can bring about.

The decline and fall of one culture is always a disaster for humanity at large, as each unique culture offers the possibility of a different way of life, paving the way for anyone who wants to embark on the adventure of attempting to live in a different cultural milieu; and thus it is the spiritual home of each individual at the highest level. It is a pity that some short- sighted social/cultural groups have caused severe cultural losses of this kind, which can never be remedied. It is fortunate that the modern world is realizing the importance of keeping alive a variety of cultures, not only for the sake of preservation, but in order to maximize partnership, cooperation, and to some extent, competition. One can surmise that a monocultural world, no matter how seemingly powerful it was, would soon suffer from decay, disruption, and finally death. An ideal model of cultural interaction is one that can highlight and help develop and spread the best aspects of both parties involved, avoiding destructive rivalry, so that humanity can grow in a healthy way, an ambition completely foreign to the one-way model of interaction.

It is certainly necessary and good for China to learn from the West. Yet the way that China has done so has been so extreme—though this is understandable considering the military and political pressure it had endured—that a one-way model was the inevitable result, leading China to undergo a process of Westernization that has thwarted healthy cultural interaction.

The uniqueness of Chinese cultural ideas and practices has resulted in many achievements that have benefited the West. Joseph Needham suggests that the introduction of Chinese science and inventions played a crucial role in the transformation of Western societies. "It is widely appreciated that their inventions changed the course of world history. Without paper, printing, the compass, and gunpowder, how would the change from feudalism to capitalism in the West have been possible?" (Needham 1969, p. 149) However, the Westernization of Chinese culture under the sway of the one-way model will inevitably cause the decline or even extinction of its uniqueness, the consequences of which will be great loss to the human future.

It is a great pity that many traditional Chinese values that might have staved off the ill effects of Westernization were dissolved by this very process. Again, Joseph Needham has written:

> What it would probably be true to say is that science in traditional China was never divorced from ethics, i.e., value-free. This kind of science was at first, in the West, highly beneficial – an unravelling of the forms of experience – but in the end it has proved most dangerous for humanity ... If mankind is not able to learn this lesson, then the utter destruction of all life on earth is inevitable. (Needham 1986, p. 21)[4]

[4] Needham's syntax isn't very clear. Yet it is clear that he meant "value-free" science has proved dangerous for humanity. Indeed, modern science has been developing so fast that it is more and more dangerous.

The science and technology that modern China has adopted from the West are causing great disasters, and a reconstruction of the balance between science and morality is urgently required.[5] In fact, the disasters caused by modern science worldwide are already posing severe threats to humanity.

There is no lack of narrow-minded and short-sighted views and deeds in the field of cultural exchange. The British and French armies in China who burned the Summer Palace in Beijing to the ground certainly did not realize what they were destroying; the vanguard of the May Fourth Movement in China who sought to eradicate the whole of Chinese cultural heritage did not fully grasp the consequences of their efforts. While politicians are professionally obligated to twist values to their advantage, scholars' ethics require them to hold to the truth, uncover and protect it, against all efforts to twist and abuse it.

The interactions of literatures, which should have distinctive features of their own, more or less have tended to follow the broader trends of cultural interaction, according to which China has learned from the West in the fields of military science, economics, and politics. While the advancement of military and scientific culture is no guarantee of literary advancement (as the example of Ancient Athens demonstrates), it is often the case that might bring about cultural superiority and influence. And yet, the beauty of literature is that its most basic literary features, such as the sound of the words and the delicate feelings expressed, are sustained by the language and culture in which they are forged. This is why literary interactions, in the free style of the two-way model alone, can enrich both parties, whereas the one-way model ends up impoverishing all. Theory, in this respect, should follow practice, developing principles of successful exchange that can be elaborated into guidelines of interaction, to which this book will make some contribution in latter chapters.

As a result of the one-way model, much of China's valuable literary heritage was not properly studied, let alone properly appreciated or inherited. Even works that benefit greatly from the traditional Chinese literary heritage, such as Xu's "Farewell Again to Cambridge," are not appropriately evaluated, because traditional Chinese literature was driven to the margins of the culture by radical and wanton innovators. The trend continues, and will certainly cause the gradual death of much valuable Chinese literary practice, which would constitute a loss to Chinese literature directly, and Western literature indirectly. Though Ezra Pound's translations from Chinese were the product of a thousand accidents and cultural crossings, it was the persistence of some perception of value in the ancient works of the Tang dynasty that kept them alive for that western poet, and his followers, to adapt them.

Both Xu Zhimo and William Empson played, almost intuitively, their part in the background of a one-way model of cultural interaction, of which they were not conscious and about which they could do nothing. Each in his own way had an admirable outcome. Yet a more conscious effort to adopt the two-way model has now

[5] Studies show that the key problem of modern China is the domination of philosophy-science, introduced into China from the West, over morality that originated in China. Thus some ways are proposed to balance them in order to relieve the many problems, of which the most serious is the decay of morality, in modern China. (Li 2011).

become possible, and it remains highly desirable, not least because the new age has revealed so many of the negative consequences of the one-way model. One promising route of interaction stems from the recent introduction of J. H. Prynne's poetry into China. The key to ongoing success in this case will be to distinguish between what should be known and what should be learned, as well as between what can be learned and what cannot: a process sure to produce many poetic experiments and adventures, and to bring about unconstrained friendship across poetic and cultural divides. And it is indeed foreseeable that a new type of two-way model of cultural interaction, different from both that of the second phase and that of the first phase, which will be characterized by absolute freedom and equality, as well as open-mindedness and the mutual devotion of the individual participants, may one day develop beyond the horizon of the current entanglements of China and the West. This book is written in the hope that some study of these possibilities may also help effectuate them.

Works Cited

Dawson, Raymond, ed. 1964. *The Legacy of China*. Oxford: Clarendon Press.

Deng, Zhongliang. 2006. Yanbusun He Zhongguo [Empson and China]. *Zhonghua Dushubao*, September 27.

Empson, William. 2000. *The Complete Poems of William Empson*, ed. John Haffenden. London: Penguin.

Gernet, Jacques. 1982. *A History of Chinese Civilization*, trans. by J. R. Foster. Cambridge: Cambridge University Press.

Haffenden, John. 2006a. *William Empson, Volume 1: Among the Mandarins*. Oxford: Oxford University Press.

Haffenden, John. 2006b. *William Empson, Volume II: Against the Christians*. Oxford: Oxford University Press.

Hao, Shanjun et al. 1996. *Zhongxi 500nian Bijiao* [A Comparison of China and the West in the Past Five Hundred Years]. Beijing: Zhongguo Gongren Chubanshe.

Hu, Shi. 1999. *Hu Shi Shuo Wenxue Bianqian* [Hu Shi's Talks on Literary Changes]. Shanghai: Shanghai Guji Chubanshe.

Li, Zhimin. 2005. *New Chinese Poetry under the Influence of Western Poetics: The Origins, Development and Sense of Nativeness*. Chongqing: Xinan Shifan Daxue Chubanshe.

Li, Zhimin. 2011. *Zhishi de Shanyuzhen* [The Good and the True of Knowledge]. Beijing: Renmin Chubanshe.

Needham, Joseph. 1969. *The Grand Titration*. London: George Allen & Unwin Ltd.

Needham, Joseph. 1986. Science, Technology, Progress and the Break-through: China as a Case Study in Human History. In *Progress in Science and its Social Conditions*, ed. Tord Ganelins. Oxford: Pergamon.

Voltaire. 1972. *Philosophical Dictionary*, ed. & trans. Theodore Besterman. London: Penguin.

Xu, Zhimo. 1927. *Balide linzhao [Scratches in Paris]*. Shanghai: Xinyue Publishing House.

Xu, Zhimo. 1983. *Xu Zhimo Xuanji* [Selected Works of Xu Zhimo], ed. Shao Huaqiang and Ying Guojing. Beijing: Renmin Wenxue Chubanshe.

Xu, Zhimo. 1990. *Xu Zhimo Shi Quanbian* [A Complete Collection of Xu Zhimo's Poems], ed. Liang Ren. Hangzhou: Zhejiang Wenyi Chubanshe.

Zenkichi, Mitsuishi. 1998. *Chuantong Zhongguode Neifaxing Fazhan* [An Endogenous Development of Traditional China], trans. Yu Xiangke. Bejing: Zhongyang Bianyi Chubanshe.

Zhang, Tongdao. 1998. *Tanxiande Fengqi: Lun 20 Shiji Zhongguo Xiandai Zhuyi Shichao* [An Exploration: On Modern Chinese Poetry Schools in the Twentieth Century]. Hefei: Anhui Jiaoyu Chubanshe.

Zhao, Xiaqiu. 1989. *Xu Zhimo Zhuan* [A Biography of Xu Zhimo]. Beijing: Zhongguo Renmin Daxue Chubanshe.

Zheng, Xiaofang. 1999. *Zhongguo Jindai Wenxuede Lishi Guiji* [The Historical Traces of Chinese Modern Literature]. Shanghai: Shanghai Shudian Chubanshe.

Chapter 3
Turbulent Transformation of the Literary and Cultural Mentality of China from 1917 to 1949 and the Vigorous One after the Reform and Opening-Up Policy

The period from 1917 to 1949 was one of the most turbulent in the modern history of China. By contrast, the Reform and Opening-Up Policy first applied in 1978 led to an era of great prosperity and relative equanimity. Each period played an important role in the transformation of the literary and cultural mentality of China. In the first period, China made a decisive turn away from its traditional literature and culture; in the second, it entered into modernity.

To gain a better understanding of these two periods, it is necessary to briefly review the modern history of China.

Endless War and Suffering in China before 1949

Wars are devastating to a nation. One Chinese saying vividly expresses how much people hate wars and love peace: "rather to be a dog in an age of peace than to be a human in a warring one." In fact, the most important reason for Chinese people in the past to harbor the imperial system and to worship a powerful emperor was that he conquered all other armed forces and thus created and kept peace. Cao Chao (155–220), who paved the way for his son to become the first emperor of Wei Dynasty (220–266), once famously said: "If I had not fought and defeated so many warlords, no one could tell how many of them had claimed themselves emperors." (Cao 1974, p. 74) It was common sense that the more emperors there were, the more wars there would be. Therefore, by claiming that he had prevented many people from proclaiming themselves emperors, Cao was boasting his great contribution to the peace of the nation, on which his very qualifications for becoming an emperor himself rested.[1]

Prof. Kyoo Lee contributed to the editing work of this chapter, which is greatly appreciated.

[1] Cao Chao did not claim himself an emperor for political reasons. After his death, his son became an emperor and established the Wei Dynasty (220–266).

Z. Li and D. Braun, *China Being Led and Leading*,
https://doi.org/10.1007/978-981-96-6803-8_3

There were often bloody wars during shifts of dynasties across the long history of China, causing death on a great scale. For example, around sixty percent of the whole population was lost from the year 184 to 220 during the collapse of the Han Dynasty (202BC–220); similarly, more than half of the population was lost during the shift from the Sui Dynasty (581–618) to the Tang Dynasty (618–907); and around forty percent was lost during the shift from the Ming Dynasty (1368–1644) to the Qing Dynasty (1616–1912). (Ge 2008, p. 94)[2] In the modern age, scholars tend to highlight the disadvantages of imperial rule. However, for the people who were actually living in ancient times, their strongest wish was undoubtedly for a powerful emperor to restore peace from seemingly perpetual wars.

In traditional China, the most fundamental political principle was probably that whoever had the most powerful army became the emperor. Therefore, military power dominated the political arena.

In 1840, Great Britain invaded China, which marked the beginning of the last stage of the Qing Dynasty. This period is also often referred to as the early dawn of modern China. The military power of Great Britain prevailed. However, Great Britain did not seek to replace the emperor, as all masters of past victorious armies had done throughout the history of China. Instead, Great Britain sought treaties that allowed it certain priorities to do business in China, which was perceived as a humiliation by the Chinese people.

The victory of the army of Great Britain demonstrated the weakness of the royal house of the Qing Dynasty, which fostered military challenges at home and abroad in the years to follow. In a certain sense, the army of Great Britain destroyed the absolute authority of the emperor, which caused the loss of many more lives than those taken in battle itself. The Qing Dynasty fell subsequently into one war after another, among which the Taiping Heavenly Kingdom Movement (1851–1864) was the most bloody, causing the deaths of between fifty to one hundred million people. (Ge 2008, p. 94) The Taiping Heavenly Kingdom Movement was the last time a military power sought to make an emperor, albeit a short-lived one, in the logic of the traditional imperial system of China. When the Qing Dynasty ended in 1912, it was replaced by a republic, headed by a president rather than an emperor, symbolizing China's entrance into the modern age in political terms, though there was still a long way to go to become well-developed for the political system, as well as the cultural system that sustains it.

The first president of modern China, which came to bear the name the Republic of China, was Sun Yat-sen (1866–1925), who was educated in the West. He drew inspiration from this education in conceiving of the republican government, modeling China's first post-imperial system on the American one. However, China was not yet ready for this transformation. Unlike George Washington (1732–1799), who willingly handed over military power to the government after his victory in the American War of Independence (1775–1783), those who held military powers in China, represented by Yuan Shikai (1859–1916), took military powers as their personal capital, holding fast to power. Indeed, Yuan Shikai succeeded in forcing Sun Yat-sen to

[2] A more detailed research work can be found in his monograph. (Ge 1991).

resign after being president for only three months, taking over as the president of the Republic of China himself. However, Yuan neither understood nor believed in the ideologies embodied in the modern political system of the republic, choosing to restore the imperial system, and he became an emperor in 1915. He was miserably mistaken in this political move, as China by then had already determined not to go back to the imperial system, though the country encountered many challenges in its effort to establish a modern political system that best suited it. Yuan died a shameful death after being emperor for only eighty-three days, which put a definitive end to any fantasy of bringing the traditional imperial system back to China – though there were still numerous cultural institutions and practices that had been co-formed in the traditional imperial system, and would take a long time to be modernized.

After the death of Yuan Shikai, China soon fell into civil wars. Whoever possessed military powers became a warlord, and different warlords fought each other, causing the death of countless people and bringing devastation to the young republic. The Kuomintang led by Sun Yat-sen was an exception in the chaos. Unlike all the other military powers, the Kuomintang was led by political creeds, which marked another big stride of China into a modern nation.

In 1927, the Kuomintang defeated the warlords and unified China. The country began to enjoy a hard-earned peace. However, the world was dominated by military powers who intended to shape and rule the world in military ways. Japan, one of the most dominant military powers of the era, invaded China and took its territory in the north-east away in 1931, and again in 1937, colonizing a large part of the whole nation. Japan's invasion tore China into pieces, and caused the death of tens of millions of civilians, until it was defeated in 1945 by the Allied countries.

After the Second World War, the USA became the most powerful nation, leading many nations to establish an international system in which all nations involved—including the defeated nations of World War II, such as Germany and Japan—could develop and prosper by making innovations, products, and trade. Wars that once were profitable started to become less so in comparison with normal economic activities in this new age, especially with respect to human cost. Therefore, the fundamental reason for waging a war disappeared. As a result, there has not been anything like the First and Second World Wars that killed millions of people up till now, though there have been some small- or middle-scaled wars, most of which brought harm instead of profit to those involved, such as the Vietnam War (1954–1975) and the Soviet–Afghan War (1979–1989). Since a war no longer brings profit, it has become more and more unpopular, and to wage a war has become a growing risk for a politician. In this sense, the military powers that were once dominant in the traditional ages have been replaced by those which bring benefits to a nation and its people, such as entrepreneurs who promote trade, scientists who innovate, statesmen who promote international business collaborations, and so on. Hopefully, the age dominated by military powers is gone forever.

In 1949, when the People's Republic of China was established, after so much turbulence and with tremendous loss of lives, China finally started to enjoy peace that has lasted until the present day, which is highly appreciated and cherished by all people in China.

Literary and Cultural Movements in China from 1917 to 1949

While there had been certain literary and cultural movements before 1917 in China, it was the Literary Revolution started by Hu Shi (1891–1962) and others in 1917 that marked the most decisive turn from tradition in literary terms. The 1919 May Fourth Movement was perhaps the most radical break in cultural terms.

The literary and cultural movements between 1917 and 1949 started with the influence of Western ideologies. The Literary Revolution was ignited by Hu Shi when he published "Some Modest Proposals for the Reform of Literature" in *New Youth* (《新青年》) in 1917. (Hu 1917, pp. 1–11) Hu Shi wrote the essay in the United States, taking many ideas from the Imagists.[3] Hu Shi, together with others in the Revolutionary Camp, achieved great success for the revolution of Chinese literature, transforming it from traditional ways to modern forms. It was the same group of people who started cultural movements afterwards, and the same strategy was applied, i.e., to learn from the West. The May Fourth Movement, arguably the most important cultural movement in modern Chinese history, took Science and Democracy, two core items from the West, as its slogan.

Nevertheless, many other forces besides Western ideologies also played their parts during the course of China's transformation from traditional to modern ways. In fact, whatever resulted was often a compromise of all forces involved, such as Western influence, the traditional cultural background, actual living circumstances, and many more, including, of course, the force of human need.

People were generally divided into two camps in the literary and cultural movements: those who supported it, including the editor of *New Youth*, Chen Duxiu (1879–1942), along with authors and readers of this popular journal, often called the Revolutionary Camp; and those who fought for Chinese literary and cultural traditions, including Mei Guangdi (1890–1945), who initiated the journal of *Xue Heng*, and the authors and readers of that journal, who were often called the Counterrevolutionary Camp. Many famous writers and scholars, as well as many successful political and military leaders, were members or followers of the Revolutionary Camp. Exemplary literary figures across China, such as Lu Xun (1881–1936), Guo Moruo (1892–1978), Mao Dun (1896–1981), Ba Jin (1904–2005), Lao She (1899–1966), Cao Yu (1910–1996), and others, produced many influential novels, plays, prose works, and poems that helped to transform the literary and cultural mentality of China from the traditional to the modern.

In what follows, we draw a more comprehensive picture of how the Chinese literary and cultural mindset was fundamentally transformed from the traditional to the modern. We outline three aspects of this transformation.

[3] Please refer to (Li 2005, pp. 42–44).

Linguistic Transformation from Wenyan to the Vernacular

In traditional China, people could be divided into two classes: the literate and the illiterate. The former comprised the ruling class, the latter, the ruled. During warring ages, an illiterate solider might become an official because of his bravery, and thus enter the ruling class, but this was rare; upon rising to the ruling class, he would ensure his children were taught literacy, thus confirming its principal importance.

There were two kinds of languages in China as well: Wenyan, or the literary language, and the vernacular, the daily spoken language. In all formal occasions, Wenyan was used—and could only be used by the literate classes. Both the literate and the illiterate classes knew the vernacular, with which they could communicate among themselves. Wenyan was more refined than the vernacular, after generations and generations of literary people's work on it, which made learning it an education in tradition as much as contemporary linguistic competence.

The most powerful stronghold for Wenyan was traditional Chinese poetry, where the most literary beauty had been created. And it was traditional Chinese poetry that Hu Shi chose to be his first target for the Literary Revolution. In 1920, Hu Shi published *Attempting Poems* (《尝试集》), the first poem-collection in vernacular Chinese. It had great influence though its literary quality was not great. All literary figures in the Revolutionary Camp followed Hu Shi in composing poetry in vernacular Chinese. As a result, new Chinese poetry, i.e., poetry written in vernacular Chinese, soon took the place of traditional Chinese poetry, becoming mainstream.[4] This was a decisive war won by the Revolutionary Camp, which firmly established vernacular Chinese as a literary language. As a result, the age of Wenyan was gone forever.

However, the power of Wenyan was undeniable; it appealed to whomever had an eye for literary beauty no matter which camp he/she belonged to. In 1928, Hu Shi published a monograph entitled *History of Vernacular Literature*, (Hu 1928) in which he made great efforts to argue that most of the beautiful traditional Chinese poems were actually composed in the vernacular. The arguments Hu Shi put forward in the book were a failure in academic terms, as Hu Shi could not produce a complete definition of the vernacular; indeed, many of the works he cited were actually written in Wenyan. His argument was, in truth, an argument against common sense, as most of the literate class would have taken all the traditional poems with strict rhythmic rules he cited in the books as being written in Wenyan. (Li 2008, pp. 129–140) However, it is understandable if it is interpreted as Hu Shi's regret for the radical attacks of the Revolutionary Camp on Wenyan and traditional Chinese literature. As he grew older and the revolutionary fervor tempered, the real intent for Hu Shi to write this book was probably to create a compromise between the vernacular movement that was meant to get ordinary people educated and the literary beauty embodied in Wenyan.

The Revolutionary Camp even went so far as to propose to abdicate Chinese Ideographic Characters as a whole and to establish another system, Pinyin Chinese, in the Roman alphabet, following Western languages. In 1918, Hu Shi claimed that "the third Word-revolution [to use Pinyin to replace Chinese characters, keeping the

[4] A more detailed review on revolution in the field of poetry is to be given in Chapter Three.

vernacular] will finally be achieved." (Hu 1999, p. 60) Hu's view was supported by many. Lü Shuxiang (1904–1998), a famous linguist in China who played an important part in producing the simplified Chinese characters that are today used on the mainland of China, insisted that "To simplify Chinese is only a temporary method. … To solve the problem forever will be the realisation of alphabetic Chinese." (Lü 1998, p. 118.) Another famous linguist, Wang Li (1900–1986), held the same view, noting, "I believe that the government will follow the will of the people and soon take measures to realize the goal to replace Chinese characters with Pinyin."(Wang et al. 1980, p. 14) The effort to get rid of Chinese characters entirely, which sounds almost unimaginable to scholars today, was indeed once quite popular. Fortunately, Chinese characters, together with the rich cultural heritage carried by it, have been kept, though simplified. Though the Chinese Pinyin system was created, it has been used as a supporting system for the character system, rather than a replacement. The current mix of systems is indeed the result of compromise, which is likely for the best.

Hu Shi believed that the Chinese language itself was inferior to Western languages, as he said that it was unable to express the sophisticated and subtle ideas carried by European languages which he had learned in studying them. In 1935, Hu Shi wrote "it is inevitable to Europeanize the vernacular Chinese, as it is the only way to meet the needs of the new age. This is to absorb the fine texture of European languages so that the vernacular Chinese could express sophisticated ideas." (Hu (I) 1998, p. 130) Such ideas were ardently supported by many famous figures in the Revolutionary Camp, such as Lu Xun, who insisted on a strategy of foreignization while translating European literary works into Chinese, even if this meant sacrificing fluency(Lu (VI) 2005, p. 77). Lu Xun noted on numerous occasions that he would rather have fidelity than fluency, and that the Chinese language, either written or spoken, was loose, and needed to be remedied by introducing the ways of European languages (Lu (IV) 2005, p. 379). Yan Fu's (1854–1921) translation work of *Evolution and Ethics* by Thomas Henry Huxley, published in 1898, testified to the contrary that Wenyan could express the most sophisticated European ideas. Yan Fu put some of his own ideas into the translation of the work on purpose. Indeed, many beautifully translated passages demonstrate that he was able to achieve the goal of carrying over into Wenyan both the beauty and sophistication of the original text. (Wang 2010, pp. 74–79).

Even though the theory Hu Shi put forward was not exact, the outcome was desirable. Modern Chinese, i.e., the vernacular Chinese, is now very much Europeanized, and it is very different from the vernacular Chinese before the Literary Revolution in 1917. Indeed, it has become very fluent after a process of refinement over a century. The most remarkable achievement in this respect was that of grammar, which was systematically introduced to Chinese by Ma Jianzhong (1845–1900) in 1898 and 1899.[5] This has made modern Chinese much more resilient than Wenyan, which has

[5] Ma Jianzhong's, *Chinese Grammar by Mr. Ma* (马氏文通) consists of ten volumes, of which the first six ones were published in 1898 and the rest four in 1899, all by The Commercial Press in Shanghai.

in fact transformed the mentality of modern Chinese people who are now able to well understand both the traditional Chinese and the Western peoples.

Supposing Hu Shi and his followers in the Revolutionary Camp had not been that radical, could vernacular Chinese have been accepted as a literary language? The answer is probably no. In this sense, the unfair treatment and damage to Wenyan and traditional Chinese literature caused by the Literary Revolution might have been unavoidable. Indeed, it is a great pity that there are only a small number of people who are specifically trained in Wenyan, especially in terms of composition, though many classic traditional literary works, of poetry especially, are still widely read.

Despite all of the chaos caused by the Literary Revolution, vernacular Chinese has been a great success. Many marvelous literary works have been produced in it, and its literary beauty has been fully demonstrated. Just as Hu Shi and his followers had expected, vernacular Chinese has indeed made great contributions to the education of the ordinary people, and has thus helped China to be much more developed.

Transformation of Literary Forms from the Traditional to the Modern

All traditional Chinese literary genres have strict formal regulations. For example, the number of lines in a poem is strictly regulated, as are the number of characters in each line, and indeed the tone of each character, with almost sonnet-like strictness. It is no exaggeration to say that traditional China was heavily regulated in all aspects of life, not only in the field of literature, but also in all fields of social life in general. The most heavily regulated was probably the officialdom, especially the royal house, where any violation of the rules might cause severe punishment or even the loss of one's life.

Both the Literary Revolution and the cultural revolution were to liberate people from all kinds of regulations. In the literary field, the greatest achievement after the Literary Revolution was in the field of novels, where famous modern Chinese writers, such as Lu Xun, Guo Moruo, Mao Dun, Ba Jin, and others, produced many influential works that molded the mentality of modern Chinese people.

In the Ming Dynasty (1368–1644) and the Qing Dynasty (1616–1912), there had been novels written, such as *Journey to the West*, *Outlaws of the Marsh*, *Romance of the Three Kingdoms*, *A Dream of Red Mansions*, and others. They were called "小说" in Chinese, or "little talk" in literal translation. The word "little" implies that they were considered inferior to such classic literary genres like poetry. Traditionally, novels were considered not to be noble, but trifling entertainments, and therefore not taken very seriously.

However, the Literary Revolution changed everything. Novels have become the most dominant literary genre in modern China, and novelists the most influential writers.

Lu Xun is undoubtedly the most important writer in modern China, who published the first novel in the modern form, "A Madman's Diary," also the first novel in vernacular Chinese, in 1918. (Lu 1918, pp. 414–424) It became an immediate success, hailed by all people in the Revolutionary Camp. Unlike Hu Shi's *Attempting Poems*, which are more revolutionary than literary, Lu Xun's "A Madman's Diary" is a literary wonder, which is still read widely. After the great success of his first novel, Lu Xun produced many more classic modern novels such as "Kong Yiji," "Medicine," and "Hometown," among others.

Traditional Chinese novels were organized in a certain style called Zhanghui, i.e., each chapter headed by a couplet giving the gist of its content. Traditional novelists paid a lot of attention to the elegance of the language, while the way they presented a story was often plain. Lu Xun never wrote in the Zhanghui style: he introduced numerous techniques for the presentation of a story, which added much literary charm to his works. For example, the story of "A Madman's Diary" is presented in the first person, i.e., a madman who speaks in a way that appears to be careless and ambiguous. However, the more a reader gives thought to the words of the madman, the more one is shocked by the truth half-revealed by these seemingly careless and ambiguous sentences. In the end, readers find that the so called "madman" is actually the only sober figure in the novel, while all others are truly crazy. Indeed, with the publication of "A Madman's Diary," a totally new field, that of the modern Chinese novel, emerged.

Following the example of Lu Xun, many writers started to compose novels in this modern form, and many wonderful works were produced, such as *Midnight* (1933) by Mao Dun, *Family* (1931) by Ba Jin, *Four Generations Under One Roof* (1944) by Lao She, *The Border Town* (1934) by Shen Congwen (1902–1988), *Xiao Erhei's Marriage* (1945) by Zhao Shuli (1906–1970), and "Lotus Creek" (1945) by Sun Li (1913–2002), to name only a few. It is difficult to calculate how many novels were produced during 1917 to 1949. However, the dozens of famous novelists and their works are well inventoried in the book *Thirty Years of Modern Chinese Literature, 1917–1949*. (Qian et al. 2024).

There were also a few novelists who insisted on writing in the traditional style, and a few of them were quite popular, as there were readers who kept the taste cultivated by traditional Chinese literature. However, their influence was much less than those in the Revolutionary Camp. They were often criticized as being blind and numb to the miserable situation of the nation and its people, who were suffering at the hands of warlords or Japanese invaders.

Modern Chinese plays and drama achieved great successes as well, second only to novels. In fact, China had a wonderful tradition of opera: the most famous varieties include Beijing Opera, Kunqu Opera, Pingju Opera, Cantonese Opera, Yu Opera, and many others. The traditional operas all had a long history of refinement, so that they were not only artistically beautiful in terms of dancing and singing, but also morally educational, as they told moving stories to promote traditional cultural values. There were many techniques and rules according to which traditional operas were to be performed, and it was a very demanding, long process to become a qualified actor or actress. In fact, most performers started to learn it in their childhood, and took acting as their lifelong profession. After the Literary Revolution, plays and dramas were introduced to China from the West and supported by the Revolutionary Camp. They soon became quite popular. To the happiness of the revolutionaries, plays and dramas were much easier to learn and to perform than traditional Chinese operas, making them the ideal platform for the promotion of education and the cultivation of the general public of China. The effect was a powerful one. For example, the so-called street drama was once very popular, in which a short play was performed by a small group of performers to pass certain ideas to the audience who were by and large illiterate. Though the drama was quite rudimentary, it was effective to get the revolutionary message out to the general public.

The most famous playwright in modern China was Cao Yu (1910–1996), who published such famous plays *The Thunderstorm* (1933), *The Sunrise* (1936), *Peking Man* (1941), and others. There are also many other famous plays, such as *Twin Flowers* (1921) and *Qu Yuan* (1942) by Guo Moruo, *The Night the Tiger Was Captured* (1924) and *Death of a Famous Actor* (1927) by Tian Han (1898–1968), to name only a few.

Great achievements had also been made in other literary genres, such as modern poetry and modern prose, which had successfully established themselves as mainstream while pushing their traditional counterparts into the background.

The Literary Revolution was undoubtedly a tremendous success. It would have been impossible even for the revolutionaries themselves to have imaged that so many wonderful literary works in the modern form could have been produced in such a short period of time, and that they would come to displace the traditional Chinese literature of thousands of years so swiftly.

Transformation of Cultural Values

The Revolutionary Camp targeted traditional Chinese values from the beginning. In 1916, Chen Duxiu published an essay that bitterly criticized traditional Chinese cultural values while singing high praise of Western cultural values. He proposed to construct modern Chinese cultural values, noting that "the last enlightenment for Chinese people, as I could see, must be that of values." (Chen 1916, p. 4) Just as they had expected and worked hard to push forwards, cultural values of the Chinese people were greatly transformed by all literary, cultural, and social movements.

The most important cultural values were those of the family in traditional China. Mencius, the most important Confucian sage next only to Confucius, said in one of the most fundamental Confucian works: "It is nations that make up the world, and families the nation." (Mencius 2021, p. 137) Qian Mu, a famous Confucian scholar in the modern age, proposed that "The family group is the most important pillar of Chinese culture, as the family group is the very foundation of the whole Chinese culture." (Qian 1994, p. 51) Qian Mu considered that the family to a Chinese person is as important as the church was to a religious Western person, noting "Chinese people get cultivated of their conscience and cultural values, such as love between father and son, or among brothers. Therefore, family to a Chinese is just like a church." (Qian 1943, p. 17) However, during and after the Literary Revolution and the Cultural Revolution, family values were once again targeted.

In many famous modern literary works in the first half of the twentieth century, the family was portrayed in a negative light. In the very first modern novel of China, "A Madman's Diary," Lu Xun fiercely attacked traditional family culture as well as the Confucianism that took it as its foundation, accusing them of a nature that "eats people," a powerfully apt metaphor for a traditional Chinese cultural system that suffocated its people (Wu 2014, p. 15). Lu Xun did not make any arguments to prove that traditional Chinese culture was suffocating people in this novel. Instead, he just showed it from the perspective of the madman himself. Readers were quickly convinced, and traditional Chinese culture was in the wake of the novel's publication often presented in the image of a cannibal, especially among people of the Revolutionary Camp.

Following Lu Xun, many other famous modern Chinese writers targeted traditional Chinese culture. For example, Ba Jin published *Family* in 1931, which told a long story of the dark rule of Confucianism in the Gao family, and the miserable fates of all the characters related to the Gao family. It was very influential and was taken as a prophecy to foretell the inevitable fate of collapse of traditional Chinese families, together with the cultural value system that backed them. *Family* by Ba Jin, *The Thunderstorm* (1933) by Cao Yu, *The Golden Cangue* (1944) by Eileen Chang (1920–1995), and many other famous literary works all revealed and criticized the dark side of the traditional Chinese family that had been considered to be an ideal, or heaven, for Chinese culture.

Traditional marriage was also severely attacked. In traditional China, the marriage of young people was taken as a key institution of the family, in which young people did not have freedom to choose whom to marry. Instead, they were arranged to get married by the leaders of their families, and their most important responsibility was to produce children, especially boys, who were to carry the family lineage on. In modern literary works such as *Family*, *The Thunderstorm*, *The Golden Cangue*, and many others, the traditional Chinese culture of marriage was portrayed as utterly inhuman. Other novels such as *Xiao Erhei's Marriage* (1943) by Zhao Shunli (1906–1970) told stories of young people who loved each other and got married by their free will, despite the fact that their parents tried all means to stop them.

The revolutionists not only criticized traditional Chinese culture in their works, but also revolted against it in real life. Xu Zhimo (1897–1931), one of the most famous modern Chinese poets, was considered to be the first person in modern China to divorce from his wife, Zhang Youyi (1900–1988), against his parents' will in 1922. In fact, almost all those who supported the literary and cultural revolutions would go on to pursue free love and free marriage. However, there were some exceptions, of which Hu Shi and Lu Xun are the most well-known. Hu Shi obeyed his parents' arrangement and got married to Jiang Dongxiu (1890–1975), an ordinary woman, and kept the marriage for all his lifetime, an ordinary marriage story shared by all traditional marriages. Lu Xun obeyed his parents and married Zhu An (1878–1947), another ordinary woman, whom Lu Xun neither slept with nor divorced. He simply ignored this wife, and married another woman, Xu Guangping (1898–1968), a college student whom he met and loved. Lu Xun's marriage reflects the most tragic cultural struggle and tension in a person during the age of cultural transformations, torn as he was, and unable to resolve the two forces of traditional and new love. (Li 2018, pp. 47–58 + 165–166.)

While the revolutionists attacked traditional Chinese cultural values, they introduced cultural values from the West as a replacement, such as freedom, equality, democracy, science, nationalism, and others, of which some took root in China more quickly and some more slowly, some with small modifications and others with a lot. Indeed, the cultural transformation of China from the traditional to the modern was something that had never happened in the thousands of years of the country's history.

Vigorous Transformation of China after the Reform and Opening-Up Policy

The Reform and Opening-Up Policy adopted by China in 1978 transformed the country in terms of its economy, which is the very soil where modern Chinese literature and culture took root. As the economic foundation determines the superstructure of a society, (Zhao 2011, pp. 57–62) the economic booming of China was to transform China in more profound ways than the literary and cultural revolutions that had taken place in the first half of the twentieth century.

The very drive for modernization for all nations, either Western or Eastern, is modern productivity, enabled by modern science and technology, and underpinned and reinforced by modern cultural and political systems. No matter how hard the revolutionaries had promoted modern cultural values in China, it was difficult for the ordinary people to truly have faith in them, so long as the economic situation of the ordinary people did not reinforce or sustain belief in those values. For example, if a person could not make a living without the support of his or her family, it would be difficult for the person to rebel against the family to claim such modern values as freedom of marriage, no matter how much one might suffer under the old regime of the family. With the agricultural way of life, most ordinary people in the countryside could not make a good living without support of their families or family groups. This is why it was so difficult for modern values to take root in the vast countryside of China before the Reform and Opening-Up Policy was introduced: for the policy was instrumental in transforming China from an agricultural country into an industrial one, which, in turn, has fundamentally transformed the literary and cultural mentality of the Chinese people once and for all. If the transformation was carried out from the top downwards during the literary and cultural movements in the first half of the twentieth century, it was from the bottom upwards that the Reform and Opening-Up Policy wrought its effects.

From an Agricultural Nation to an Industrial One

China had been an agricultural country for thousands of years, with a well-established system of cultural institutions that reinforced this agricultural modality. The Chinese scholar Feng Tianyu (1942–2023) has said that it was because of the agricultural lifestyle that the traditional Chinese culture of families was both invented and inherited. (Feng 1994, p. 6) However, when China met the West in the modern age and saw the power of the products of modern industry, it started to turn to industry as well. In 1861, the government of the Qing Dynasty issued an order that initiated the so-called Westernization Movement, an effort that was meant to learn from the West to build up modern industries in China to increase its economic power. However, the movement was considered a failure because the Qing Dynasty was utterly defeated in the Sino-Japanese War of 1894–1895, though it did introduce modern technology and industries into China from the West, and began the dream of industrialization in China.

After the Westernization Movement, modern technology and industries continued to develop in China, but at a rather slow pace, in part because wars of aggression continued to wrack the country. In 1949, after peace was secured in China, the Chinese government ardently promoted the development of modern industries and, indeed, a whole system of industries, together with its supporting system of modern science and technology. Despite this progress, China was still a predominantly agricultural nation, with more than 82.08 percent of its population being farmers before the Reform and Opening-Up Policy.[6]

The whole economic landscape of China started to shift rapidly after the Reform and Opening-Up Policy was applied in 1978, when capital and technology started to flow into the country at an accelerated rate from developed nations, mainly those of the West. Within a few decades, China became the so-called world factory, and was the second largest economy in the world by 2010. Indeed, after more than forty years of fast economic growth, China became an industrial nation. In 2020, people who were living in rural areas had dropped from 82.08 percent in 1978 to only 36.1 percent. The population continues to flow from rural to urban areas, suggesting that this trend of urbanization will continue in the years to come.[7] Many people who remain in rural areas are senior citizens, or those who cannot find a good job in cities. Undoubtedly, China has become an industrial nation, which means it has finally entered into modernity more generally.

From a historical perspective, this could be taken as the realization of a dream — or at least a large part of it—of the people who launched the literary and cultural

[6] Cui Hongyan and Li Guizhi, "Changzhu Renkou Chengzhenhualv Ruhe Cesuan" ["How to Estimate Urbanization Ratio on Basis of Resident Population in Urban Areas"], on website of *National Bureau of Statistics* of China, posted on January 1, 2023.
See https://www.stats.gov.cn/zs/tjws/tjzb/202301/t20230101_1903783.html.

[7] Cui Zhenwu, "Diqici Quanguo Renkou Pucha Gongbao Jiedu" ["Explication on Statistics of the Seventh National Population Census"], on website of *National Bureau of Statistics*, of China, posted on May 12, 2021.
See https://www.stats.gov.cn/sj/sjjd/202302/t20230202_1896484.html.

movements more than a century ago. China has embarked on a modern way of life, and there is no way to go back, though there are surely a lot of difficulties waiting ahead for Chinese people to overcome in the future.

Modern Values Getting Rooted in China

When the revolutionists urged their followers in China to actually practice the values imported from the West in the early twentieth century, it caused more harm than good to their followers, as society was not yet ready for such values. Lu Xun published a novel called "Regret for the Past" in 1925, telling a tragic story of a young couple who fought for free love and free marriage yet eventually failed, the wife committing suicide, leaving the husband desperate and aimless. With this story, Lu Xun warned young people who were fascinated with modern values that they might risk their lives if they followed them heedlessly. In the countryside, where most residents were farmers, modern ideas ardently proposed by revolutionaries sounded impractical. If one dared to put them into practice, one was sure to be punished, perhaps even killed.

Tragedies such as the one depicted in "Regret for the Past" by Lu Xun have nothing to do with the modern values themselves. Rather, the tragedy was one of untimeliness, a function of the values produced to serve a modern industrial society that did not comprehend a still largely agricultural nation. However, the situation was different following China's industrialization. Today, modern values have become a way of life for more and more Chinese people, whether they have learned them purposefully or understood them well or not.

When the West transformed from the traditional age to the modern, the most profound change was that of people's attitude to the church, which had been the center of their lives. In China, it was the family. Since most Chinese are no longer leading an agricultural lifestyle, the traditional Confucian culture exerts less and less influence upon those working and living in cities far from their families, the network in which the traditional cultural values has held fast. Indeed, in the traditional agricultural society, a person is first and foremost a family person. In the modern industrial society, a person is first and foremost a citizen of the society, who has the most direct contact with the society itself, in the form of a job and necessary facilities to make a living.

In the modern industrial society, everybody who could find a job could become economically independent, and free to marry, without dependence on their families for arranging nuptials, and without fearing consequences like those depicted in Lu Xun's "Regret for the Past." Therefore, people became independent and began to enjoy the freedoms of modern society. Indeed, in the modern way of life, people intuitively took to modern values that granted them a better life. In this way, modern values began to take root in China in a way impossible to walk back from.

Conclusion

To transform from the traditional to the modern is never easy. It was difficult for the West, and it has been difficult for China, especially since China is such a large nation with a long history.

It was the West that pushed China into the modern world. Many scholars believe that China might never have changed its traditional way of life if it had not been because of the constant push of the West, both with military invasion, and in the role of a model of prosperity. (Qian 1996, p. 167) It is also because of the dual sides of the West that were revealed to China that it has been both criticized and admired at the same time. The more modern China becomes, the more modern values it shares with the West, the more it will be able to play the role as a friend and a partner, and the more contribution it will make to the modern world.

The West itself was brought into the modern age because of the development of modern philosophy, science, and technology, which is the real drive of modernization for both the West and China. Despite the fact that it is the West that pushed China onto the road of modernization, it is modern philosophy, science, and technology rather than the West itself that are the true drive for China's modernization. The West serves as a successful model to China in many respects, such as economic achievement, while an awkward example in other aspects, such as racial harmony.

China time and again attempted the lazy way to modernize itself by copying from the West, and failed and suffered as a consequence, which testified that China had to work out its own way to modernization. China could and should learn from the West, but not copy its ways blindly.

Despite all of the hardships, stumbles, and suffering, China has achieved great successes towards modernization, such as those achieved in the Literary Revolution, cultural revolution, and economic reformation. It has been now most widely agreed by the whole population that China must continue its journey to modernization until its maturation.

Works Cited

Cao, Chao. 1974. *Cao Chao ji* [Collected Works of Cao Chao]. Beijing: Zhonghua Shuju.

Chen, Duxiu. 1916. Wuren Zuihouzhi Juewu [The Last Conscience to Be Wakened-up for Chinese People]. *Qingnian Zazhi [youth Journal]* 1 (6): 1–4.

Feng, Tianyu. 1994. *Zhongguo Wenhua Shigang* [On History of Chinese Culture]. Beijing: Beijing Yuyan Daxue Chubanshe.

Ge, Jianxiong. 1991. *Zhongguo Renkou Fazhanshi* [History of Chinese Population]. Fuzhou: Fujian Renmin Chubanshe.

Ge, Jianxiong. 2008. Zhongguo Lidai Renkou Shuliangde Yanbianji Zengjiande Yuanyin [The Changes of Population and Their Causes in Different Dynasties in China]. *Dangde Wenxian [literature of Chinese Communist Party]* 2: 94–95.

Hu, Shi. 1917. Wenxue Gailiang Chuyi [Some Modest Proposals for the Reform of Literature]. *Xinqingnian [new Youth]* 2 (5): 1–11.

Hu, Shi. 1928. *Baihua Wenxueshi* [History of Vernacular Literature]. Shanghai: Xinyue Shudian.
Hu, Shi. 1998. Hu Shi Wenji (1) (2) [Collected Works of Hu Shi (I) (II)]. Beijing: Beijing Daxue Chubanshe.
Hu, Shi. 1999. *Hu Shi Shuowenxue Bianqian* [Hu Shi's Talks on Literary Changes]. Shanghai: Shanghai Guji Chubanshe.
Li, Zhimin. 2005. *New Chinese Poetry under the Influence of Western Poetics: The Origins, Development and Sense of Nativeness.* Chongqing: Xinan Shifandaxue Chubanshe.
Li, Zhimin. 2008. *Shixue Goujian: Yixiangyu Xingshi* [Poetics Constructed: The Form and the Image]. Beijing: Renmin Chubanshe.
Li, Zhimin. 2018. Lun Lu Xunde Ziwo Wenhua Duizhi: Zouchu Chuantong-xiandaide Wenhuamiju [On Lu Xun's Inner Cultural Tension: Getting out of the Cultural Puzzle of Tradition vs. Modernity]. *Wenshizhe* [Journal of Literature, History and Philosophy] 2: 47–58+165–166.
Lu, Xun. 1918. Kuangren Riji [A Madman's Diary]. *Xinqingnian [new Youth]* 4 (5): 414–424.
Lu, Xun. 2005. *Lu Xun Quanji* (4) (6) [The Complete Works of Lu Xun, (IV) (VI)]. Beijing: Renmin Wenxue Chubanshe.
Lü, Shuxiang. 1998. *Yuwen Changtan* [Common Talks on Chinese Language]. Beijing: Sanlian Shudian.
Ma, Jianzhong. 1898–1899. *Mashi Wentong* (1–10) [Chinese Grammar by Mr. Ma (I-X)]. Shanghai: Shangwu Yinshuguan.
Mencius. 2021. *Mengzi Yizhu* [Mencius with Translation and Annotation]. Yang Bojun and Yang Fengbin. trans and annotated. Changsha: Yuelu Shushe.
Qian, Liqun et al. 2024. *Zhongguo Xiandaiwenxue Sanshinian* [Thirty Years of Modern Chinese Literature, 1917–1949]. Beijing: Beijing Daxue Chubanshe.
Qian, Mansu. 1996. Aimosheng Hezhongguo: Duigerenzhuyi Defansi [Emerson and China: A Retrospect of Individualism]. Beijing: Sanlian Shudian.
Mu., and Qian. 1943. Kongziyu Xinjiao [Confucius and Conscience Education]. *Sixiangyu Shidai* [Thoughts and Eras] 21: 15–18.
Qian, Mu. 1994. *Zhongguo Wenhuashi Daolun* [A Short History of Chinese Culture]. Beijing: Shangwu Yinshuguan.
Wang, Dongfeng. 2010. Tianyanlun Yiwen Pianduan Shangxi [Analysis of Paragraphs in Yan Fu's Chinese Translation of *Evolution and Ethics and Other Essays*]. *Zhongguo Fanyi [chinese Translators Journal]* 5: 74–79.
Wang, Li et al. 1980. Cujin Gaodengyuanxiao Wenzigaigede Jiaoxuehe Keyangongzuo (shang) [To Promote Teaching and Studying of Reformation of Chinese Characters in High Education (I)] *Yuwen Xiandaihua* [Modernization of the Chinese Language] 1: 1–39.
Wu, Yiqin, et al. 2014. Chiren Xushide Lishi Bianxingji [Variations of Human-Eating Narratives in History]. *Wenyi Yanjiu [literature & Art Studies]* 4: 14–22.
Zhao, Jiaxiang. 2011. Jingjijichu Jueding Shangcengjianzhu Yuanlide Xingcheng Guochengji Xitonglunzheng [Formation and Systematic Testification of the Theory of the Economic Foundation Determining the Superstructure of a Society]. *Beijing Xingzheng Xueyuan Xuebao [journal of Beijing Administrative College]* 1: 57–62.

Chapter 4
The Origin and Development of New Chinese Poetry from the Perspective of Cultural Exchanges between China and the West

In this age of globalization, no nation can survive, not to say develop, without learning from the achievements, either of natural science or of social science, of other nations. "Nations must serve as guides for one another." In every country one should "welcome foreign ideas, for hospitality in this way makes the fortune of those who receive it" (Wellek 1981, p. 230). China is fortunate in this regard as it has been keen on receiving influences in the modern age from the West who had received influences from China in previous ages.

Influence of China on the West

Joseph Needham once said: "The more you know of Chinese technology in the medieval period, the more you realize that, not only in the case of certain things very well known, such as the invention of gunpowder, the invention of paper, printing, and the magnetic compass, but in many other cases, inventions and technological discoveries were made in China which changed the course of Western civilization, and indeed that of the whole world" (Needham 1969, pp. 149–154). Needham offers many concrete examples to support his argument. Natural science is itself an embodiment of a certain philosophy. And indeed to accept the science and technology of another culture is to accept the philosophical point of view which underwrites those disciplines and developments as well. The great literary critic Hugh Kenner has traced numerous examples of Chinese philosophic influence on Western philosophy (Kenner 1972, p. 231). In the field of politics, basic institutions, including the "Civil Service Examination System," were directly imported from China to Europe. As we have already seen, Literature also made its way through select translations from China into the West, exerting a powerful influence on the sensibilities of intellectual and artistic circles across the centuries following the initial missions to eastern kingdom in the sixteenth century. Johann Wolfgang von Goethe, as is well known, was full

Z. Li and D. Braun, *China Being Led and Leading*,
https://doi.org/10.1007/978-981-96-6803-8_4

of praise for the wisdom of Chinese literature, singling out the novel *Haoqiu Zhuan* (《好逑传》), *The Fortunate Union*, as having contributed to his concept of world literature.

In the field of poetry in particular, China may be said to have borne the greatest influence on western practitioners. Perhaps the most well-known example of this influence is the case of the twentieth century modernist poet, Ezra Pound whose *Cathay*, a book of lyric poems freely translated from Chinese sources (with one exception) became an international sensation upon its publication in 1915. "The River-Merchant's Wife: a Letter" from this volume exemplifies the charm of Chinese poetry that attracted Pound. The whole poem is full of fresh, concrete, energetic images, which are created by the technique of "Look-Back Strikes." For example, in lines 2–4, an ordinary poet might just write: "I played about the front gate/ You came by on bamboo stilts/ You walked about my seat," in which the story of the romance at the center of the poem's longing is also made clear. However, the poetic beauty of the images is terribly undermined by a certain abstract quality. The three phrases "pulling flowers," "playing horse," "playing with blue plum" serve as "Back Strikes," hammering once more on the images which precede them and guarantee their beauty. Together, three phrases produce a marvelous poetic rhythm and a series of sound images as well. In the image conjured by the lines "By the gate now, the moss is grown, the different mosses,/ Too deep to clear them away," two "Back Strikes" are applied, producing a vivid image. However, the most marvelous poetic achievement is the unprecedented, and probably subconscious, application of "Sound-Linked Images Flow" in English. This poetic technique is very frequently applied in traditional Chinese poetry, and especially obvious in Chinese Couplets. It is a technique which frees images from the bondage of grammar (without becoming agrammatical) in order to link them on a higher plane of semiosis; the sound of words closely knits them into a beautiful acoustic unit. In "The River-Merchant's Wife: A Letter", the phrase-images "pulling flowers," "playing horse," "playing with blue plum," though not typical examples, more or less achieve this effect.

In fact, *Cathay* was such a success that, when it came out in 1915, "almost all the major Modernists—Yeats, Ford, Lewis, Eliot, and Williams—applauded its freshness, elegance, and simplicity. Ford, in particular, remarked, 'If these were original verses, then Pound was the greatest poet of the day,' which should indicate that the English Modernist writer and critic had perceived affinities between his own and Far Eastern sensibilities" (Qian 1995, pp. 3–4). These keen figures might have perceived something of this renovation through sound-image. Following Pound, Western poets began to learn more and more from Chinese poetry. William Carlos Williams, his friend and frequent correspondent in these matters, spoke for example, of "the technique of achieving wholeness by means of uniting opposing elements," which he had learned from Bai Juyi, a contemporary of Li Bai from the Tang Dynasty period (Qian 1995, p. 137).

New Chinese Poetry Under the Influence of the West

Though readers are doubtless familiar with the examples of the influence exerted on western modernism by its discovery of ancient Chinese poetry, it remains surprising that this current of cultural exchange in fact ran both ways. It is no understatement to say that New Chinese Poetry was born with the introduction of Western Poetry. In order to clarify the nature of this influence, we must first examine the social, cultural and literary backgrounds in China under which New Chinese Poetry was conceived and born.

China had not known much of the West until the year of 1840, when Great Britain waged a war against the Qing Dynasty government, a campaign which would eventually lead to the dawn of China's modern age. China had remained a powerful nation for a long time. Even at the time when Columbus discovered America in 1492, China was still the most powerful nation in the world. "The most powerful empire the world over was in Asia—she was China in Ming Dynasty. From 1405 to 1423, Chinese fleets which contained hundreds of warships dwarfed Columbus's sailboats and brought the fame of the Chinese Empire further into South Asia, East Asia and the Indonesian islands" (Chua-Eoan 1991, p. 7). However, the West made very rapid progress after the Renaissance, and eventually surpassed China.

In 1840, China failed in the war with Great Britain, a failure followed by a series of misfortunes she had never experienced in the past. As a result, "The Chinese began to doubt anything of their traditional culture and this gradually caused the May-fourth movement that was to overthrow all traditional Chinese cultures and rebuild their culture by learning everything from the West civilizations ..." (Pulleyblank 1964, pp. 80–81). Throughout its history, China had been learning from other nations, accepting various forms of cultural nourishment (Gernet 1982, p. 20). Yet the process by which China came to learn from the West after 1840 was unique. "When China was confronting the challenge of the West in the modern era, Confucianism could no more absorb it as it once had done with Buddhism. It finally became the negative force in the development of Chinese society, causing the Chinese to fall into a severe ideological peril" (Qian 1996, p. 3). Confucianism, once the main foundation of Chinese culture, suddenly appeared deficient and was radically criticized by segments of the Chinese cultural vanguard, especially those who had spent time studying in the West. Chinese culture then experienced a collapse from within, resulting in an overwhelming influx of Western ideologies. Chinese literature, as a main bearer of Chinese culture, was first and foremost challenged by these new ideologies. Even the Chinese ideographic characters that were established thousands of years prior to the Opium Wars, characters which had been the medium of the creation of many literary wonders, were suddenly questioned and challenged as to their aesthetic suitability (Hu 1999, p. 60). It was a chaotic age, one marked by the questioning and re-examination of the roots and branches alike of Chinese cultural values. In the vacuum of authority, Western models flooded in, and in many cases rapidly replaced traditional cultural practices.

Poetry, as the most sensitive element of Chinese culture, responded very fast to the dramatic changes following the year 1840. Not long after the conclusion of the Opium War, initial reformations could be observed. Gong Zizhen (龚自珍) and Wei Yuan (魏源), two famous poets, for example, made the patriotic claim that there was an urgent need to reform Traditional Chinese Poetry in order to serve the nation (Zheng 1999, p. 61). Both made important contributions to this effort; but it was Huang Zunxian (黄遵宪) and Liang Qichao (梁启超) made a true beginning at rendering classical Chinese poetry into a freer form style profoundly influenced by Western poetics.

The most well-known resulting poetic reformation during this period was Huang's "New Style," the core principle of which is sometimes summarized by the phrase: "to write as you speak." Liang, however, cultivated a much grander aim for the reformation of traditional Chinese literature, looking at the whole of the system and setting his challenge to traditional literary language itself, or Wenyan. Later on, people took Liang Qichao as the initiator of Baihua Literature (literature in vernacular Chinese), and many important literary figures, including Hu Shi (胡适), Chen Duxiu (陈独秀) and Li Dazhao (李大钊), were deeply influenced by Liang before the May Fourth Movement.

Changes in the field of poetry only serve as one example. In fact, changes occurred across all literary fields following the Opium War, guided by the basic belief that Chinese culture needed to reform in order to survive (Zheng 1999, pp. 4–5). These reform movements exploded in 1919, culminating in the May Fourth Movement, which strenuously championed the Western ideologies of Democracy and Science. Its fundamental significance was its acceptance of the need to learn from Western culture in reforming Chinese traditional culture and literature. Hu Shi's "Some Suggestions for Reforming Chinese Literature" (Hu 1929, p. 7), which was composed in the United States and posted back to China to be published by Chen Duxiu in *New Youth* (《新青年》), the vanguard magazine of the May Fourth Movement, was a significant first salvo in the bourgeoning culture war, and was enthusiastically hailed and followed. Hu Shi advocated the establishment of a New Chinese Literature by the overthrow of the traditional art, and the traditional literary language in which it was written, Wenyan.

In place of traditional language and poetics, Hu Shi argued that vernacular Chinese was in fact the true bearer of poetry, and to rid themselves of traditional habits and ideas, poets must embrace this language of the people, as it was spoken. "It is very clear that the vernacular could be applied to writing poems. The vernacular poems of Du Fu (杜甫), Bai Juyi (白居易), Hanshan (寒山), Shi De (拾得), Shao Yong (邵雍), Wang Anshi (王安石) and Lu You (陆游) can all be cited as proofs. There are even more proofs in Ci (词) and Qu (曲)" (Hu 1999, p. 205). Adding, in sum (and exaggeration), that "in short, all valuable literary works in China since *The Book of Poetry* (《诗经》) are in the vernacular or very close to the vernacular" (Hu 1999, p. 45). Hu Shi's criteria were of course partial. Nevertheless, his standpoint was supported by many followers, who agreed to simplify traditional Chinese literature, eliminating the Wenyan of the educated classes, so that the ordinary public could access their writing, and educate themselves to the May Fourth program for the culture of the nation.[1]

Many scholars admit that the creation of New Chinese Literature came about mainly because of the need to call on all Chinese to come together to save their nation. I. A. Richards, remarking on his own time spent in China, said: "The intellectual movement in Modern China is primarily a consequence of the political movement. The traditional Chinese outlook is being remade—not because it was felt to be unsatisfactory in *itself* but because it plainly put China at a disadvantage in the world-struggle" (Richards 1932, p. 102). Since traditional Chinese poetry appealed only to a small number of well-educated people, it was criticized and replaced by easily understandable New Chinese Poetry. Naturally, that did not mean that traditional Chinese literature had not been beautiful and nourishing to the spirit of those who were not persuaded by the new poetic ethos that swept the country (nor those who had previously studied and appreciated the traditional works). Literary beauty and literary popularity are not identical; in fact, they are each internally heterogeneous, and heterochronic, shifting in their meanings and reach with the vicissitudes of history and taste. Sometimes the two coincide as aesthetic priorities, but more often they diverge.

Traditional Chinese Poetry had undergone several large-scale changes in its history before the arrival of China's modern age. "After Four-word poetry declined, there appeared *Songs from the South* (楚词); after *Songs from the South* declined, there appeared Five-word poetry; after Five-word poetry was Seven-word poetry; after the ancient-style poetry was Lu (律) and Jue (绝), followed by Ci (词)" (Wang 1999, p. 86). Wang Guowei (王国维) gave a more academically reasoned explanation for this phenomenon: "...all those changes happened because after a certain literary form got into circulation, more people would utilize it and then a rigid pattern would take shape, which would prevent even the talented from making attractive literature. Therefore, people abdicated the old forms and created new ones. This is the reason that all prosperous literary forms finally declined" (Wang 1999, p. 86). However, Wang's explanation cannot quite explain the formation of New Chinese Poetry, because the movement and its ideals had been established and widely accepted even before any literary works were created to justify the genre. In fact, the ideas may have more successfully rooted themselves in the consciousness of readers than the works themselves. Throughout the twentieth century, many figures, including poet Chairman Mao Zedong (although he supported New Chinese Poetry politically), repeatedly questioned the success of New Chinese Poetry as a literary genre.

[1] It is ironic to note that Hu Shi and his colleagues worked hard to simplify Chinese poetry, however, after more than one century's development, contemporary Chinese poetry is becoming more and more difficult again, being mostly too difficult for ordinary readers to understand. Hu Shi could have been more convincing if he had not sought too exclusively poetical reasons rather than social ones. Hu Shi and his fellow Cultural Reformers did succeed in establishing vernacular Chinese as an officially accepted written language. However, if it had not been for the particular historical background, Hu Shi would never have succeeded so easily in establishing the vernacular, instead of Wenyan, as the solely right tool for poetry, because his arguments would never have been strong enough to do so in any other historical situation.

The Formation Period of New Chinese Poetry from the 1910s to the 1940s

A desperate person can easily be driven into madness, so can a people. Since Chinese culture seemed too weak to hold its own against the violence of internal warfare and external forces of modernization, it was subsequently challenged in more and less insane ways. Chinese classical literature, being a part of Chinese culture, with Chinese classical poetry as the central component, was supplanted by a so-called commonplace people's literature, the main aim of which was to penetrate into the minds of the ordinary Chinese, calling upon them to stand up for the rescue of their nation and themselves.

During those years, many who were best positioned to interpret Western culture and literature—usually, those who had spent significant periods abroad—acquired a considerable cultural influence. Hu Shi, who had stayed for eight years in the United States and finally earned a Ph.D. at Columbia University in New York, was among the most influential of these figures. It would be no exaggeration to state that he "ignited" the New Chinese Literary Movements. Hu wrote: "This movement [the May Fourth Movement]—which is generally called a literary revolution, I myself prefer to address it as the Chinese Renaissance—was initiated by my friends and I in an American university dormitory in 1915, 1916 and 1917" (Hu 1999, p. 259). Hu Shi's Doctorate was not in literature but in ancient Chinese philosophy, though he claimed expertise in Western literature as a result of his time in the United States – though the extent of his knowledge and reading has more recently been questioned by scholars.

Hu Shi's *Attempting Poems* (《尝试集》), the first poem-collection in Vernacular Chinese, which was widely distributed and imitated, formally launching the New Chinese Poetry, is actually of quite limited literary quality. Its most significant value seems to consist in the very fact that it is in vernacular Chinese—like "a small foot that has just been released" as Hu Shi himself later on apologized. It is almost certain that Hu Shi indeed learned or copied some models from Western sources in constructing his famous "Eight Nos,"[2] a prescription which he drew up for reforming Chinese literature, and which became highly influential during subsequent literary movements in China (Hu 1921, p. 7). In fact, while Hu Shi was studying in New York from 1910 to 1917, he encountered the manifestos and poetry of Imagism, which were being promulgated by Ezra Pound and Amy Lowell. Indeed, Hu copied portions of their various proposals (and famous "Don'ts" from 1911–1914) into his notebook (Pan 1991, p. 29). The Imagists proposed to create free verse in new-rhythms and new-forms using English vernacular, rather than high-flown literary language. There can hardly be any doubt that Hu borrowed some ideas from the Imagists, who had themselves learned from Traditional Chinese Poetry (Zhang 1998, pp. 105–106).

[2] They are: (1) whatever said should mean; (2) not to imitate the ancient; (3) pay attention to grammar; (4) no false emotion; (5) no cliché; (6) no idioms; (7) no antithesis; (8) not to avoid the vulgar language.

Hu Shi, in a sense, is one of the few persons who set the literary direction for the May Fourth Movement. His Western background and Western literary characteristics thus strongly promoted the Western poetical characteristics of New Chinese Poetry. Soon after Hu's *Attempting Poems*, there emerged many collections that more or less based themselves on imported Western poetics, among which Guo Moruo's (郭沫若) *The Goddesses* (《女神》, 1921) and Li Jinfa's (李金发) *Light Rain* (《微雨》, 1925), which drew their inspiration and form respectively from Romanticism and Symbolism, received the most immediate and widespread attention in Chinese literary circles.

Guo Moruo's poem *The Goddesses,* which is generally considered to mark the very beginning of Chinese Modern Romanticism, the first climax of New Chinese Poetry, is perhaps the most obvious in its Western borrowings (Huang 1997, pp. 81–94). The title image "goddess" is from Western mythology. "The Western images in *The Goddess* are much more prevalent than the Chinese ones, such Western mythic images as Apollo, Venus, Cupid, Bacchus, Prometheus, Hygeia… There are also many Western historical terms. And there are even many words in English in this Chinese poem" (Wen 1995, pp. 406–407). Western words had never been directly employed in traditional Chinese poetry, and it is unclear in what sense a poem composed of many Western images that only the educated elite could understand was written for the common Guo claimed as his audience. This founding irony has in fact been a consistent characteristic of New Chinese Poetry, which continued to adopt Western poetical developments as it evolved, rendering it less and less accessible to common readers.

As soon as *The Goddesses* was published, it quickly became popular within Chinese literary circles, firmly establishing Guo's position as a leading figure on the literary scene. However, the most important factor in the appeal of *The Goddesses* was not so much its literary quality, as rather its patriotic themes. The collection as a whole enthusiastically hailed China as a phoenix, risen from the ashes of cultural destruction. The theme of triumph is most apparent in the poem "Rebirth of the Goddesses": "… Ladies and Gentlemen, you have become tired of living in the foetid gloom of this dark world. You surely thirst for light. Your poet, having dramatized so far, writes no more. He has, in fact, fled beyond the sea to create new light and heat. Ladies and gentlemen, do you await the appearance of a new sun? You are bid to create it for yourselves. We will meet again under the new sun" (Guo 1978, p. 7). Guo's *The Goddesses* is remarkably optimistic, conveying the requisite mix of pride and ambition that were precious in an age of dark humiliation. Most of the other poems in *The Goddesses* are happy and energetic, such as "Sunrise," "Good Morning," "Shouting on the Rim of the World, Oh Earth, My Mother," "Sea of Light," "Hymn to the Sun" and "Stirring of Spring," balancing this projection of optimism with a celebration of the present.

It is often the case that literary popularity and literary merit are not correlated with one another. In *The Goddesses* there are so many superfluous words and unqualified sentences that a well-trained eye could clearly see Guo's lack of precise control over diction and emotion, such as in the poem "Good Morning":

Ah, Ah! Alas the Pacific Ocean!
Good morning! Alas the Pacific Ocean!
Good morning! Alas the New Continent beside the Pacific Ocean!
Good morning! Alas the grave of Washington! Alas the grave of Lincoln!
Alas the grave of Whitman!
Ah, Ah! Alas Whitman! Alas Whitman! Alas the Pacific-like Whitman!
Ah, Ah! Alas the Pacific Ocean!
Good morning! Alas the Pacific Ocean! (Guo 1998 , p. 66).

Any emotion, not to say unqualified diction, that exceeds the control of the poem's governing design should be considered superfluous and superficial, apparent to trained readers if often lost on the laity. It is no wonder that there should have been so many superfluous and shallow poems intended only to "serve" the common audience that were produced during the Cultural Revolution Years, while Guo held the official leadership in the literary circles in China. Guo may have understood his own poetic limitations, or come to understand them, having noted frankly that "Viewed as literature, these poems may disappoint the reader. Let them rather be taken as recordings of the age in which they were written" (Guo 1978, p. 1). It should also be remembered here that to point out the few drawbacks in *The Goddesses* is by no means to deny its significance as a literary milestone in the history of New Chinese Poetry, nor to denigrate Guo the poet as one of the greatest literary figures in the history of New Chinese Literature. Indeed, Guo also wrote several remarkable novels that deserve far wider recognition.

Li Jinfa, who introduced Western Symbolism into New Chinese Poetry, might be one of the few figures to become a famous new Chinese poet without mastery of either Chinese or any other language. Li, who was concentrating on studying sculpture in France, occasionally scratched a few lines in the manner of French Symbolism, was able to publish these poems in China to immediate and resounding success (Chen 1998, pp. 53–56). Li's introduction of French Symbolism into China attracted the attention of many poets, such as Dai Wangshu, Wang Duqing, Bian Zhilin and others, many of them now considered better writers than Li himself.

Commonly referred to as a "Shi Guai" (诗怪, a bizarre poet) because of his "bizarre" disposition of images, the name is fitting for Li, since his bizarre images themselves and the bizarrely awkward diction that made his poems appear to be very fresh and imaginative resound across his corpus. Readers of Li's works often find incoherent pictures which yield a wonderful poetical experience; nevertheless, the work also demands much imaginative power in its readers. While Li's work did win poetical merit, much of the innovation he produced had the character of accident, and it is generally agreed that he was no master of the Chinese language. This is in part circumstantial: Li had few opportunities to improve his Chinese while young (Chen 1998, pp. 29–148). This linguistic limitation is clearly on display in all three of his poetry collections, *Light Rain* (1925), *I Sang for Happiness* (《为幸福而歌》, 1926), *The Long-Term Visitor and Hard Times* (《食客与凶年》, 1927), which are marred by diction that cannot be excused as innovation. Zhu Ziqing once wrote of his work: "There is no want of imagination in Li's poems; but his sentence structures are too Europeanized, with some classical Chinese preps and conjunctions, making one feel rather like reading a piece of translation work than an original; this might be owing to Li's too anxious intention of creating a new language, or because his mother tongue is too limited" (Zhu 1983, p. 99).

Following the publication of *The Long-Term Visitor and Hard Times* Li wrote not another line in the style of his previous work. He was not alone in this abandonment of the movement he had helped to spark. Many famous writers, such as Lu Xun and Zhou Zuoren did exactly the same: they wrote New Chinese Poems in order to help establish it, only to forsake the movement and its literature soon after it had risen to prominence. Hu Shi, perhaps the poet who played the most influential role in the founding of the movement, largely did the same. For literary revolutions and their works, born of the necessity of their moment and crafted according to the demands of the new constituencies they are seeking to form, often dissolve just as quickly as they spring up. This is not entirely the case with New Chinese Poetry, in so far as it lives on in its influences, yet the production of explicit work devoted to the advance of the movement and its priorities, vanished not long after it succeeded in capturing the literary market.

The Misty Poetry Period from 1978 to 1989 and the Individual Creation Period after 1990s

New Chinese Poetry during the period from 1949 to 1978 is to be only briefly discussed, for New Chinese Literature as a whole did not make much progress during this period, especially during the Cultural Revolution from 1966 to 1976. The stagnation was caused in large part by the efforts to make literature serve as a subordinate instrument for politics, which resulted in the extreme formalization and generalization of all literary creation in China: The "revolutionary" personages in most "revolutionary" writings are perfect in all respects, the former chapters of such works informing readers about the latter ones. Therefore, "during all the ten years of the Cultural Revolution, there occurred only eight Model Plays and one Author, as people said in joking tones during those days" (Zhang 1997, p. 189). The creativity of literary circles, including those of poetry, was depressed as a result of these formal strictures. By the late 1970s, the slogan "literature serving as a subordinate instrument for politics" was officially abolished, thus liberating literature from its chains. Misty Poetry gradually grew, and finally bloomed.

It is conventional wisdom to say that Misty Poetry is made up of the sighs, complaints and songs of long depressed, and newly released intellectual spirits. We should not find it strange that Misty Poetry was mainly concerned with political themes, and was criticized as Political Poetry by some critics. More than politics, however, Misty poetry was also concerned with humanity, cultural tradition and intellectual life, being much more advanced in comparison with the poetry in the previous age.

Misty Poetry as a whole is a misty term, shrouded in a sometimes deliberate seeming vagueness. The Misty poet par excellence, Gu Cheng, once said: "Some of my fellow poets and I have always considered the term 'Misty' itself to be rather misty" (Gu 1990, p. 168). Misty Poetry was perhaps considered vague because most readers, especially some of the most influential critics of the period, had long grown accustomed to the easy, slogan-like poems which dominated New Chinese Poetry. The misty quality of Misty Poetry yielded profound poetic beauty, establishing an embryo of a new Chinese literary language that was independent of both Wenyan, the classical poetical language, and Baihua, the Vernacular Chinese that had been more or less formalized by the previous generation of literary works. Where the poetic imagination is at ease, it is often at its best. A person without the refined literary sensibility was likely to be confused and troubled, confronting a "misty" poem, while one trained on western poetics of deliberate vagueness was likely to feel inspired and gratified. This dichotomy of sensibility informed the response to the movement.

Although most Misty poets spent little to no time abroad, their works were nevertheless immediately recognized as Western influenced. "The best example to indicate the similarity between New Chinese Poetry and Western poetry is Misty Poetry, which appeared in China during the 1970s" (Lu 1991, p. 7). Shu Ting, who was judged to be following Western poetics to a great extent, was one of the most popular Misty Poets. "The majority of critical discussions believe that Shu Ting's poems are mainly influenced by the West… Some papers state that the poetical techniques of modern European Romanticism, Symbolism and Imagism can be found in Shu Ting's most excellent works… Other papers declare that, concerning the matters of Form, Theme and Methods of Expression, Shu Ting deliberately imitated the Western Poets…" (Ding 1996, p. 64). In fact Misty Poetry was characterized by the use of diverse Western techniques.

Compared with their predecessors in the first half of the twentieth century, who had visited and studied in the West for extended periods of time, many Misty Poets accepted aspects of Western poetics that came to them already translated. The major disadvantage of this is that they had unavoidably missed a lot of poetic beauty of the original works, but on the flip side, they accepted Western influence in an inwardly digested way. In other words, they accepted a wider range of what they believed would be beneficial for their poetical practice, even to the extent of intuitive acquisition, which prevented them from adopting something that was not consistent with the Chinese language or culture.

The years around 1990 are generally considered to be the end of the poetical period of Misty Poetry, and the beginning of the new period of the1990s, sometimes referred to as "a new age of Individual Creation" (Yang 1999, p. 3). The so-called Individual Creation tendency, which appeared in embryo during the Shenshen Poetry Exhibition of 1986, refers to the fact that every poet was coming to be seen as writing in his or her own particular way. The number of schools, under this new dispensation, seemed almost to be equal to the number of poets. This may have owed a great deal to both the freer social atmosphere and the great success of the Misty Poets (millions of copies of their works were sold out). The proliferation of poets and schools was on prominent display during the exhibition held by the *Shenzhen Youth Newspaper* and *Poetry Newspaper* in 1986. Around eighty different poets' schools and poetical dogmas were presented during this exhibition, with the total number of schools in China at the time believed to have been more than double that number. As the many –isms and schools were making many experiments, radical or moderate, of which some were failures, others successes, the poetical characteristic of the era of the1990s, an age of no poetical authority, gradually took its form. The many schools and –isms made the pro-classification critics feel very confused, dizzy and desperate. "Post-Misty poetry is a kind of blend of New Classicism, New Romanticism, Modernism, etc. None of them can be classified into any certain school" (Zhang 1998, p. 571).

The date that Haizi (海子), a prominent new Chinese poet who killed himself on the Shanhai Guan railway over Easter in 1989, chose for his suicide would prove a sign of further Western features to emerge within the group of New Chinese poets writing into the 1990s. In a book entitled as *The Selections of the Jottings of Contemporary Chinese Poets*, all the thirty-five poets who contributed their work can be observed frequently quoting foreign poets, critics, philosophers or experts. Few make mention of any Chinese poet or critic, either classical or modern. Another collection, *A Poem-Selection of the Most Popular Chinese Poets in the 1990s,* featured poets such as Xi Chuan (西川), Bai Hua (柏桦), Yi Hai (义海), Lin Dongwei (林东威), Mo Yaping (莫雅平), BC-1 and others who had graduated from Foreign Language Departments in Universities in China. Beyond these two prominent collections were still other poets of the New Chinese Poetry movement with specific ties to the West. Wang Jiaxin (王家新), for example, stayed in England for two years and travelled in Europe widely; Lu De-an (吕德安) stayed in the United States from 1991 to 1994; Liang Xiaoming (梁晓明) claimed to have begun writing poems after he had been greatly moved by reading Walt Whitman's *Leaves of Grass*; Hei Dachun's (黑大春) proclaimed his favorite works to be Rabindranath Tagore's *Stray Birds* and the lyrics of Lord Byron and Alexander Pushkin; Ma Li's (马莉) literary creeds were Aestheticism and Mysticism; Nie Pei (聂沛) said the first book he read was Ernest Hemingway's *The Old Man and the Sea*, while the first poet he became familiar with was William Shakespeare.

One particular poetry group, the Chinese Language-Poetry Group declared at the very beginning of their manifesto: "We object to localist and ethnocentric—in a word, to nationalistic writings. Our writings are of the world (an open box): through writing, we hold conversation with contemporary art of global significance and with modern Western art in particular" (Che et al. 1994, p. 98). This manifesto was published in 1988, ten years after Bruce Andrews and Charles Bernstein initiated their journal $L = A = N = G = U = A = G = E$, in February 1978. However, according to Zhang Ziqing (张子清), a leading figure who specializes in American Literature in China, American Language Poetry was not introduced into China until the summer of 1990 (Bernstein et al. 1993, p. 4). It is not important whether the Chinese Language-Poetry Group was shaped under the influence of their American counterpart or not; the fact that they are so eager to "hold conversation with the West" speaks to the power of simultaneity and the larger invisible forces driving poetic invention in remote yet often secretly linked areas of the creative world.

The following poem "Flowers of Two Persons", written by the leading figure of the small Chinese Language-Poetry Group Che Qianzi (车前子), might shed more light on this issue of poetic development:

Flowers in two lips, grow,
Some red words, like two roses.
Two roses,
One is taller than the other, above blue light,
The slightly shorter one leans.
Against its lover's shoulders, murmuring;
China pink, golden;
Flowers blooming in eyes, for rooms at night,
Pave carpet, purple meridian and parallel lines;
The planet suspended between two bodies,
Blooming more violets than sea water.
One warship of violet,
Another warship of violet, gold-spot-jumping.
Sunlight, the flower centre of violets, the intersects in.
The carpet, like two persons with hands crossed.
Embracing for the past.
The gold timepieces in the tender wool of the purple carpet,
As if having China pink at hand, she seizes time.
Two persons, having relationship with flowers,
Having been years. At the intersects of.
Growing and blooming: embracing for brains. (Che 2000, p. 69)

To the majority of Chinese readers, this poem likely makes no sense. It is more foreign than foreign—it is a habit among Chinese readers to describe anything difficult to understand as foreign. Yet there indeed exists a magic beauty in the work of strange juxtapositions. The perception of this beauty, however, has less to do with any culturally specific program of aesthetics or training of readers, and everything to do with the power of imagination, defined more than two centuries ago, by Samuel Taylor Coleridge: that is, to recognize and appreciate beauty created from seemingly unrelated and incoherent images, without declining into the work of mere low-grade fancy (Coleridge's term merely associative and pleasing, rather than integrative and deeply creative, work of the mind) (Coleridge 1993, pp. 295–306).

Conclusion: New Chinese Poetry as a Melting Pot for Traditional Chinese Poetry and Western Poetry

In traditional China, poetry enjoyed an extremely high social status. Almost all emperors in China's history were trained to read, appreciate, and compose poetry. Some of them were good poets; one outstanding example is Li Yu (李煜, 937–978),

who composed many wonderful works. Mao Zedong ought to be considered one of the best poets China produced during the twentieth century, though he composed only in traditional Chinese poetic forms. In the traditional age, but to an extent following the period of modernization as well, the skillful poet was a revered figure in Chinese society. Poetry was considered the beating heart of Chinese culture, indeed, its very spirit.

The difference of the cultural and social backgrounds between Chinese and Western poetry granted each very different spirits and traditions. In China, poetry was generally of the ruling classes and most poets were a part of them, and Western poetry was more of the people and more poets were plain persons; thus Chinese poetry was mostly for the ruling class, having an aristocratic tendency while Western poetry was more of the people, having a very strong rebellious tradition. John Milton in England is a good representative of the rebellious poetical spirit, while in China, whatever poets were allowed to do is at most to complain about social conditions. If a poet dared to denounce, or call for rebellion against social conditions in their poems, they might be severely punished, even to be put into jail. However, the spirit of Chinese poetical tradition changed dramatically in the early twentieth century, when New Chinese Poetry took form under the direct influence of the West. Drawing on this heritage, contemporary Chinese poets are now among the most rebellious spirits in China, perhaps in the world.

Because of the social difference between the situations of Chinese and Western poetry, Chinese poetry was concerned more with aesthetics and Western poetry with ideologies, though this is perhaps better stated as a difference of degree in each category, than a distinction according to kind. However, after the introduction of Western poetry into China, New Chinese Poetry re-founded itself as a vehicle for ideological work, taking up social and political issues nearly to the exclusion of all else. In fact, even the famous Misty Poetry Movement was so political that some critics referred to Misty Poetry as itself a Political Poetry.

Yet the strong appetite for beauty is exactly the same in both Chinese and Western poetics; this is probably why a beautiful Chinese poem can always find readers in the West and vice versa. "Inasmuch as we are all members of one human race, Chinese culture is our culture. Poetry is the heart of Chinese culture. The heart of Chinese poetry beats in us, too" (Whincup 1987, p. vii). This convincingly demonstrates that Chinese poetry and Western poetry can indeed learn and benefit from each other. In fact, foreign culture has long been a rich source for Chinese poetry. "The only other force that has had as much effect on Chinese poetry as the folk song is foreign literature. As far back as the Han Dynasty (202 B.C.-A.D. 220), military tunes from the non-Chinese tribes in the north and Northwest enriched Chinese poetry. And throughout the centuries, foreign verse forms and influences, like the indigenous folk song, have repeatedly brought fresh blood into the main artery of Chinese poetry" (Hsu 1963, p. xii). Nevertheless, the possibility of productive interchange does not mean that the approach can be applied blindly. It is clear that attempts at importing Western poetics themselves, if highly commendable, require much care in their development and execution, lest they degenerate into mere theatre of invention.

Learning is good, but only by digestion, and followed with reasonable creation. Qi Baishi (齐白石), a famous Chinese painter, once said to his admirers: "It is good to learn from me, yet it is fatal to copy me" (Lu 1991, p. 371). "Never to copy" is probably a good first rule in any transcultural exchange in the field of arts. Throughout Chinese history, there has been no lack of successful experiences of learning from foreign cultures. The acceptance of foreign influence very often contributes to the development of Chinese literature and language. Even the Four Tones of Chinese language were shaped under foreign influence. Chen Yinke (陈寅恪) in his "Four Questions about the Four Tones" has written, "the Four Tones were created by some Chinese monks and scholars who knew music well in Qi and Liang of Nan Dynasty, being inspired by Xiyu [it refers to Ancient India and the nations around it in those days] monks' ways of reading Jing [the Buddhist books]" (Hu 1998, pp. 39–40). There has been a very strong voice in the poet's circle insisting that the best way to ensure the prosperity of Chinese poetry is to "blend Chinese and Western Culture together," even as early as in the late nineteenth century. "The giant of Poetic Revolution, Huang Zunxuan, had much the same views as Liang Qichao, i.e., they both considered that the way out for Chinese poetry is to blend Chinese and Western culture together" (Zheng 1999, p. 16). However, before any blending, the following few points must be addressed: (a) a good understanding of oneself must be attained; (b) a good understanding of the other must be forged; (c) a good understanding of differences and similarities must be grasped; (d) a good understanding of the positive and negative consequences of exchanges must be canvassed. And yet most important, and perhaps least quantifiable, is that the conduit of this exchange should be the genius of poetry.

There did and does exist a chance for New Chinese Poets to learn from both Traditional Chinese Poetry and Western poetry in these ways. "In June of 1923, in the paper of 'The Local Colour of *The Goddess*', Wen Yiduo said: 'I am always thinking that New Chinese Poetry should be different from not only Traditional Chinese Poetry, but also from Western poetry. That is to say: New Chinese Poetry should not remain purely local, yet it should keep its local character; it should not be totally Western, yet it should absorb Western poetic advantages. New Chinese Poetry should be the son produced by a marriage of Chinese and Western poetics'" (Tang 1996, p. 194). Some famous Western scholars have also stated their expectation of combining the advantages of Chinese and Western cultures together. No less than I. A. Richards wrote:

> We have not yet realised that the quality of the world-life that our children are to inherit depends quite as much upon the character of the new China as upon any other one factor we can think of. If we were wiser we would make incomparably greater efforts to see that the values we most believe in came to the Chinese as freely and by as good channels as we could devise. … Only those who have had the opportunity of working with Chinese students know how great their powers and how needlessly entangling and frustrating their difficulties are. A new age is being created in this wedding of East and West. Those to whom history gives some faith in humanity will envy the Chinese the richness of the joint heritage which will be theirs. (Richards 1932, p. 103)

However, no matter how strongly people like to blend Chinese and Western cultures with poetry as their beating heart, this marriage is certainly not like one between two lovely persons in a warm church. No dream could be realized without due efforts!

Works Cited

Bernstein, Charles. Hank Lazer and James Sherry. 1993. *Selected Language Poems*, trans. Zhang Ziqing and Huang Yunte. Chengdu: Sichuan Wenyi Chubanshe.
Che, Qianzi, et al. 1994. *Original: Chinese language-Poetry Group*, trans. Jeff Twitchell. Brighton: Parataxis Editions.
Che, Qianzi. 2000. *Shikan* [Poetry Journal] (7).
Chen, Houcheng. 1998. *Li Jinfa Huiyilu* [A Memorization of Li Jinfa]. Shanghai: Dongfang Chuban Zhongxin.
Chua-Eoan, Howard. 1991. Empires on the Wane. *The Times*, October 21.
Coleridge, Samuel Taylor. 1993. *The Collected Works of Samuel Taylor Coleridge. Vol. 7 Biographia Literaria*, ed. James Engell and W. Jackson Bate. Princeton: Princeton University Press.
Ding, Guocheng. ed. 1996. *Zhongguo Xinshiqi Zhengmingshi Jingxuan* [A Selection of the Contending Poems Published in the New Era of China]. Changchun: Shidai Wenyi Chubanshe.
Gernet, Jacques. 1982. *A History of Chinese Civilization*, trans. J. R. Foster. Cambridge: Cambridge University Press.
Guo, Mo-Jo [Guo Moruo]. 1978. *Selected Poems from The Goddesses*, trans. John Lester and A.C. Barnes. Beijing: Foreign Languages Press.
Guo, Moruo. 1998. *Nv Shen* [The Goddesses]. Beijing: Renmin Wenxue Chubanshe.
Gu, Cheng. 1990. *Selected Poems by Gu Cheng*, ed. Sean Golden and Chu Chiyu, trans. John Cayley et al. Hong Kong: The Chinese University of Hong Kong Press.
Hsu, Kai-yu. 1963. *Twentieth Century Chinese Poetry*. New York: Anchor Books.
Hu, Shi. 1921. *Hu Shi Wencun (Diyice)* [Hu Shi's Work Collections (Volume 1)] (Shanghai: Shanghai Dongya Chubanshe.
Hu, Shi. 1929. *Hu Shi Wenji* [Hu Shi's Works]. Shanghai: Shanghai Dongya Chubanshe.
Hu, Shi. 1999. *Hu Shi Shuo Wenxue Bianqian* [Hu Shi's Talks on Literary Changes]. Shanghai: Shanghai Guji Chubanshe.
Hu, Yingjian. 1998. *Dushang Gaolou. Chenyinge* [Alone in a High Building: Chen Yinge]. Jinan: Shandong Huabao Chubanshe.
Huang, Xiuji. 1997. *Zhongguo Xiandai Wenxue Fazhanshi* [The History of the development of Modern Chinese Literature]. Beijing: Zhongguo Qingnian Chubanshe.
Kenner, Hugh. 1972. *The Pound Era*. London: Faber and Faber.
Lu Jin, 1991. *Zhongguo Xiandai Shixue* [Chinese Modern Poetics]. Chongqing: Chongqing Chubanshe.
Needham, Joseph. 1969. *The Grand Titration: Science and Society in East and West*. London: George Allen & Unwin.
Pan, Songde. ed. 1991. *Zhongguo Xiandai Shilun Sishijia* [Forty Poetical Schools in Modern Chinese Poetry]. Chongqing: Chongqing Chubanshe.
Pulleyblank, E. G. 1964. Chapter 3 Literature. In *The Legacy of China*, ed. Raymond Dawson, 80–164. Oxford: Clarendon Press.
Qian, Mansu. 1996. *Aimosheng he Zhongguo: Dui Geren Zhuyi de Fansi* [Emerson and China: A Retrospect of Individualism]. Beijing: Sanlian Shudian.
Qian, Zhaoming. 1995. *Orientalism and Modernism: The Legacy of China in Pound and Williams*. Durham: Duke University Press.
Richards, I. A. 1932. The Chinese Renaissance. *Scrutiny* 1(2).

Tang, Hongdi. 1996. *Shiren Wen Yiduo de Shijie* [The World of the Poet Wen Yiduo]. Shanghai: Xuelin Chubanshe.
Wang, Guowei. 1999. *Renjian Cihua* [Comments on Human Poetry], trans. Teng Xianhui. Changchun: Jilin Literary History Publishing House.
Wang, Jianzhao. ed. 1998. *Zhongguo Dangdai Shiren Suibixuan* [The Jottings of Contemporary Chinese Poets]. Beijing: Zhongguo Shehui Kexue Chubanshe.
Wellek, René. 1981. *A History of Modern Criticism: 1750–1950*, vol. 2. Cambridge: Cambridge University Press.
Wen, Yiduo. 1995. *Wen Yiduo Shi Quanbian* [Wen Yiduo's Poems]. Lan. Dizhi. Hangzhou: Zhejiang Wenyi Chubanshe.
Whincup, Greg. 1987. *The Heart of Chinese Poetry*. New York: Anchor Press.
Yang, Ke. ed.1999. *Jiushi Niandai Shili Shiren Shixuan* [Selected Poems of the Most Popular Poets in 1990s]. Guilin: Lijiang Chubanshe.
Zhang, Tongdao. 1998. *Lun 20 Shiji Zhongguo Xiandai Zhuyi Shichao* [An Exploration: On Modern Chinese Poetry Schools in the Twentieth Century]. Hefei: Anhui Jiaoyu Chubanshe.
Zhang, Dexiang, 1997. *Xianshi Zhuyi Dangdai Liubianshi* [The History of the Changes of Realism in Modern Times]. Bcijing: Shehui Kexue Wenxian Chubanshe.
Zheng, Xiaofang. 1999. *Zhongguo Jindai Wenxuede Lishi Guiji* [The Historical Traces of Chinese Modern Literature]. Shanghai: Shanghai Shudian Chubanshe.
Zhu, Ziqing. 1983. *Zhu Ziqing Xuba Shupingji* [The Prefaces and Comments Written by Zhu Ziqing]. Beijing: Sanlian Shudian.

Chapter 5
The Interplay of "Common Morality" and "Philosophy-Science"

The ancient Chinese developed their first system of moral values around the twenty first century BC, following progress in agriculture and husbandry that allowed them to meet their material needs. The story "Yu the Great Controlling Water" is among the first cultural artifacts to represent this progress. In the process of comprehending and regulating their collective lives, the ancient Chinese gradually developed a system of shared moral beliefs and practices that took human relations as its core: the final flowering of this process is the system of Confucianism.

Confucianism as a System of Common Moralities

The key to a culture lies in its moral system. (Gan 2006, p. 26) It is generally believed that Confucius (551 BC–479 BC) created the school of Confucianism. Or to put it accurately, the main contribution of Confucius himself was his systematization of the inheritance of the long tradition of moral values developed from antiquity. Qian Mu has written, "compared to the claim that Confucianism (together with Confucian thought) prescribed Chinese culture, it is fairer to say that Confucianism is a natural product evolving out of the cultural elements of antique society of China." (Qian 1994, p. 65) In agreement with Qian Mu, Feng Tianyu clarifies that "the school of Confucianism inherited the *wu* (Shaman) culture of the Yin and Shang dynasties, and developed the tradition of rituals in Western Zhou." (Feng 1994, p. 40) He goes on to conclude that Confucian culture is grounded in "the hierarchical preference structure embedded in the kinship structure," a product of "the patriarchal clan system of blood ties," which "originated in the Shang Dynasty and finalized in Western Zhou." It is

This chapter was published as "The Interplay of 'Secular Morality' and 'Philosophy-Science,'" in *NBU Journal of Philosophy* in March of 2011. Zheng Jie provided kind help in translating the essay from Chinese into English.

Z. Li and D. Braun, *China Being Led and Leading*,
https://doi.org/10.1007/978-981-96-6803-8_5

due to the fact "the majority of ancient Chinese adopted a farming life in groups," he adds, that "the clan system and its remnants based on blood tie survived many transformations that Chinese social structure underwent." (ibid., p. 6) Instead of inventing Confucianism, Confucius recorded, advocated and transmitted a system of shared beliefs and practices by means of which ancient Chinese structured and regulated their collective lives to meet the needs of society. For example, the central concept of Confucian teaching, *li* (ritual) "derives from the customs and practices of the primitive society." (ibid., p. 32) The contribution of Confucianism to the Chinese moral system lies in its codification, classification and systemization of the ethical values already largely shared by the common people.

The vitality of Confucianism lies in the fact that it is essentially a kind of Common Morality. By Common Morality we refer to a system of moral beliefs which are solely invented, agreed upon, and practiced by a broad group of people within a certain social community. Because of their basis in common Chinese tradition and belief, the wide acceptance of Confucianism was not achieved by force. While the Chinese emperor is theoretically recognized as the ruler of "all under heaven" (i.e., the world), in practice, the supreme imperial power in most cases loses jurisdiction at the local level: in other words, most rural areas are in the state of self-governance. If commoners were not willing to abide by a school of moral values, no one could really force them to practice it. Confucianism was widely accepted because commoners liked it. One important reason for the popularity of Confucianism lies in the fact that it endorses a system of morality based on "mutual benefits" (or "reciprocity"). For example, the teaching of "kind father, filial son" implies a bi-directional condition: the notion of the "kind father" encourages parents to care for their children, while the corresponding notion of the "filial son" prescribes that children should serve their parents when required to. The equilibrium of benefits enables everyone to be rewarded according to their contribution, first at the level of the family, and in larger concentric circles of social responsibility that ripple outward from it. Nevertheless, Confucianism has come under attack from many different directions from the early modern history of China. Certain Confucian teachings are amended and improved to accord with social developments.

The moral system of Confucianism is the basis of social stability in China. Among the most significant features of the system that contributes to this stability is the Confucian conception of moderate desire—as embodied in this line: "In practicing the rule of propriety, a natural ease is to be prized." (Chapter 12 The *Analects*)—which promotes an art of peace, held in high esteem by all practitioners. Confucians assume the whole of society can live and work in peace and contentment on the condition that each person observes moderation in all things. This emphasis on moderation not only expresses the people's wish for an ideal state of orderliness and peace, but also serves to govern the common people's conduct effectively, and in a large degree strengthen social cohesion and hence, promote social progress.

The Spring and Autumn Period and the Warring States Period (722–221 BC) witnessed the cultural flowering of "One Hundred Schools of Thought." However, it was Confucianism that gained prominence over other doctrines and was elevated to official orthodoxy during the reign of Emperor Wudi (141 BC–86 BC). It should be noted that "The Huang Lao Strategy" (a mixture of Legalist and Taoist thought), which was elevated as the dominant ideology of the imperial court by Emperor Wudi's forefathers, was also effective in certain ways. However, Confucianism was much more powerful to unite the whole nation as a system of common moralities, so that it was chosen by Emperor Wudi.

Taoism, another inheritor of traditional Chinese culture, is full of ancient wisdom. By contrast to the simple and clear moral teachings of Confucianism, Taoism is based on an approach consistent with dialectical monism. Its central message—"The Tao in its regular course does nothing (for the sake of doing it), and so there is nothing which it does not do" (Chapter thirty seven *Tao Te Ching*)—is not widely understood and accepted among common folks. Taoist thought is more appealing to scholars, who devote themselves to profound reflection. We may distinguish the teaching of Taoism from Confucianism in terms of "the ethics of self-cultivation" it encourages, rather than the Common Morality of the latter. The school of Legalism inherits ancient political wisdom, transmitted in legalistic policies that could be adopted (and were adopted in Qin Dynasty) to establish effective laws and procedures in state governance. However, Legalism is a kind of mix of political theory and ethics rather than Common Morality.

Following Wudi, the works of Confucius were made the official imperial philosophy. In order to win over public support, rulers demonstrated the validity of "the rule by virtue" by encouraging moral education of the masses, bestowing generous rewards on the filial sons and loyal women of the lower class. Confucianism was elevated as the dominant ideology by the rulers of different dynasties to serve the practical purpose of supporting and safeguarding their political power. Xunzi once said: "The ruler is the boat, the common people are the water. It is the water that bears up the boat but also the water that capsizes it" (Chapter 9 *Xunzi*). The historical lesson enciphered in this metaphor is that any ruler who turns his back on Common Morality will bring social disturbance, and risk the ultimate collapse of his rule.

Christianity as a System of Common Moralities

The ancient Greeks did not have a system of common moralities that was observed and practiced by all Greeks. The Greeks were religious, thus their moral values were closely bound up with their religions and embodied in their religious activities. Ancient Greek religions were polytheistic: each Greek city-state contained and worshiped its own divinities and practiced its own rituals pertaining to its pantheon. While religion served a socially cohesive function within each city-state, Greek culture as a whole lacked unity and consistency of religious and moral outlook.

Though certain features of a universal Greek culture did exist, the identification of the Greek citizen with the whole of such a culture was weak to non-existent. Viewed from within the city-states, there was no commonly accepted morality that bound all levels of society together. Greek gods displayed the features of an ideal type of behavior—which, surprisingly, centered around power instead of goodness—that the believers projected upon them. In the epic poetry of Homer, for example, gods showed interest only in interacting with heroes, kings and nobles. It is evident that while the divinities were believed to intervene in human affairs, they had little to no truck with the common people, to say nothing of the slaves who, in some areas of ancient Greece, accounted for 31–50% of the population (Liu and Wang 1994, pp. 267–268; Xu 2000, pp. 109–117). No agreed upon principles for moral conduct could be said to govern all Greeks. The ancient Greeks did not have a system of common morality. Instead, they had a kind of morality of the noble classes that was not meant for ordinary citizens, but rather served as a remote ideal for merely a small, and largely imaginary portion of the population.

The philosophers Plato and Aristotle studied these traditional ethics embodied in the literature of ancient Greece and devised their own moral philosophies, each of which centered around four cardinal virtues: temperance, courage, wisdom and justice. However, these ethical standards do not prescribe any moral rules or demonstrate any concern for actions and consequences. In other words, these ethical standards are unobtainable and thus unpracticable for the vast majority of Greek citizens: indeed, the goals of these virtue philosophies are virtually impossible to achieve for any but the heroic and philosophical class. The impractical nature of these ethical standards also meant that common people had no guidance or means for achieving authentic self-fulfillment through ethical practice. By contrast, the goals of Common Morality are designed to be achieved in daily life, and furthermore, equally apply to people of all social classes—regardless of their birth, fortune and social ranks. The lack of anything like a Common Morality (not only between the city-states but also within the city-states) in ancient Greece proved damaging in the long run to social cohesion and stability, eventually leading to the collapse of the ancient Greek world.

Christianity demonstrated its features as a system of common moralities when it appeared as a religion built upon the foundations of Judaism. Judaism is the religion of the Jewish people and is observed by all Jews, which makes it obvious that Judaism is a form of Common Morality. Because Judaism is a system of common moralities, it is of tenacious vitality and claims a historical continuity spanning more than 3000 years into the present day. Meanwhile, as it is open only to Jewish people, it serves as Common Morality only to them. However, Christianity, whose main roots are in Judaism, is open to all people who would follow it. In fact, it is often portrayed as the religion of the poor and the slaves; the description owes much to its initial popularity among slaves and the poorer members of society, and its so-called inversion of the Greek paradigm of noble and hero worship. The Christian belief—that all the followers of Christianity theoretically become sisters and brothers regardless of their fortune and social ranks actually—confers on its members an equality in the eyes of the moral law that is controverted by their social circumstances.

In modern times, it was Christians who led the fight in the West against slavery. Slave systems appeared in the early history of the United States. The motives of the abolitionists who sought to destroy them ran a broad spectrum: among the most dominant explanations in the scholarship is an economic one: i.e., the Southern economy and military effort depended on slave labor, while the rapid spread of manufacturing and commerce in the North demanded free labors available in the market. An unpaid Southern workforce had damaging effects on wages and prices in the North. (Jiang 2005, pp. 59–61) Following this logic, the emancipation of slaves would be beneficial to the industrial development of the North. Yet the limitation of the theory is obvious: if the Southern economy was attacked with a heavy blow due to the abolition of slavery, the economic development of the North would be affected as well. We should be reminded that when Abraham Lincoln met Harriet Beecher Stowe, the author of *Uncle Tom's Cabin*, at the start of the Civil War, he declared, "So this is the little lady who made this big war." At risk of provoking hostile responses from those Americans who supported slavery, Stowe published her novel asserting Christian love's triumph over the reality of slavery. Stowe intended Uncle Tom, a noble but long-suffering Christian slave, to be a "noble hero." The image of Uncle Tom was so overwhelming among the white Christian readers that antislavery sentiments were aroused: slavery was denounced as a sin that must be abolished. Therefore, Christianity as Common Morality played the most fundamental role in igniting the Civil War that would eventually bring about the end of chattel slavery in America.

Although slavery in the US was abolished in 1865, racial discrimination and segregation persisted. In the twentieth century, the leaders of the American civil rights movement demonstrated a distinctively Protestant character in their outlook and founding political philosophy for American democracy and culture. Among them, Martin Luther King jr., the civil rights activist and pastor, in his famous "I Have a Dream" speech, frequently referred to God and claimed that American people—either white or black—are "all of God's children," in order to emphasize the Christian roots of American spiritual values and stress the moral imperatives to end sinful practices. The connection of civil rights movements and American politics more broadly to these values is abundantly evident to this day. Barack Obama, in his election campaign, emphasized his religious background. Indeed, it is clear that Christianity played a crucial role in abolitionism, and in promoting social progress and maintaining social stability.[1]

[1] Joseph Needham believes that China did not ever pass through a "slave society" analogous to that of classical Greece and Rome, "a slave-based on society as the Mediterranean culture" (Needham 1969, p. 201). Needham's claim is made from the economic and political perspective. We could tackle this issue from the ethical perspective. To understand whether "chattel slavery" was ever a basis for the whole society in China as the same way as it was in classical Greece and Rome, we need to question in the first place what it means to be a slave in the Mediterranean culture. In ancient Rome, slaves were sometimes sent to entertain audiences in violent confrontations with wild animals in the arena. Drawn from the fact that Romans experienced "aesthetic satisfaction" from the gladiator games, we can conclude that apparently slaves were treated as the equivalent of animals. In other words, Roman ethics were not applicable among slaves since they were stripped of their citizenship. For those who argue that China did pass a "slave society," their main argument is as follows: Since the emperor is the only authorized governor of his people, thus all people are his slaves, as evinced in this teaching: "If the king decrees death to his subject, his subject will do as the kings says." Following this logic we could claim that all sons are slaves of their fathers because of another teaching: "If the father decrees death to his son, his son will do as the father says," which is absolute nonsense taken literally. From the ethical perspective, we realize that in the Chinese context there never appeared the galley-slave pattern characteristic of Mediterranean Europe. Joseph Needham used "kuan-liao" to refer to the conception of Asian bureaucratism; we

The ultimate goal of religion is directed towards life. This may seem a controversial statement, yet the Christian God, while promising an afterlife in heaven for believers, focuses more on how to live the present life. As Liu Xiaofeng remarked: "The prophets in the Old Testament instead intend to direct people's attention to the earthly life itself." (Liu 2007b, pp. 230–231) Christians find in the *Bible* the most fundamental guidance on morality and human behavior in daily life, including the guiding virtues of diligence, honesty, gratitude, fraternity, forgiveness, and so on. The Ten Commandments as the moral foundation of the Hebrew *Bible* is a list of imperatives, which, according to the *Bible*, were spoken by God to Moses, who gave them to the people of Israel. The commandments are seen as the commonly accepted moral imperatives, among which the first three statements are of religious faith and the following seven are basic values that the believers are supposed to abide by. The seven statements are as follows: 4. Diligence (Labor for six days and rest on the seventh day—the Sabbath day to the God); 5. Honor your father and your mother; 6. You shall not murder. 7. You shall not commit adultery; 8. You shall not steal. 9. You shall not bear false witness against your neighbor. 10. You shall not covet anything that belongs to your neighbor. The first three statements (concerning the relationship between God and humans)—which establishes faith (faithfulness) and belief (obedience) in God—serve as a sound basis for the rest of the commandments: they ensure the absolute validity of the other seven statements as a requirement binding upon Christians and, furthermore, ensure the consensus of moral values among the common people. Here we can see that Christian morality and Confucian morality, albeit varied in their theoretical foundations, display shared characteristics as Common Morality and a disposition toward social cohesion and stability in the notion of a single code of conduct for all believers (citizens).

Reformation of Christianity

The vitality of Christianity as Common Morality was demonstrated in the beginning of its formation. This new cult swept over the Roman Empire like a tidal wave, which caused great fear among the Roman authorities who carried out ruthless persecutions of early believers in response. Persecution of the Christians began with Nero's bloody repression of 60 AD, and became more severe during the second and third centuries. The fact that many Christians were mercilessly murdered did not stop the growth of Christianity, but rather had the opposite effect, ensuring it spread further across the empire and its social strata. Realizing that repression was not effective, the Roman government had to change its tactics. At a meeting of Bishops in Milan (313 AD) the so-called Edict of Milan was passed which essentially recognized the legitimization of all religions including Christianity. In 380 AD, Christianity became the official religion of the Roman Empire and ultimately captured the whole Roman Empire.

---

might as well use "斯勒夫" to refer to the Western conception of the institution of slavery, as we could not find an equal term in the Chinese language.

Christianity as Common Morality could survive and develop under severe repression. Meanwhile, it could sponsor deadly movements that had many people killed. For instance, the Crusades (a series of religiously sanctioned military campaigns) that caused many deaths were waged in the name of a restoration of Christian control over the Holy land. In modern times, with "sword in one hand, and the *Bible* in the other hand," European countries intruded into the rest of the world and established colonies on other continents. From the beginning of the twentieth century onwards, with the decline of political and military forces of European countries, they slowly lost their control of colonies and hence withdrew their military force from the former colonies. Nevertheless, Christianity that was spread through missionary work and colonization was kept in many previous colonies. Indeed, historical events demonstrate that the power of Common Morality is mightier and more persistent than any kind of military force. The tremendous empire conquered by Alexander, which was established solely upon its military force and spread across the continents of Asia, Europe and Africa, collapsed and was split so quickly after his death. The same fate befell the great Yuan Empire founded by Kublai Khan. Such cases prove that military force, no matter how powerful it seems to be, is often short lived.

From the time of its commencement, Christianity as a secret organization was dedicated to building a Christian society within the existing framework of the Roman Empire. Its characteristics of being subversive to Roman law and order required strict organization in Christian society to keep conformity between thought and action among its members. Therefore, the early Christians with their strong monotheistic convictions did not tolerate other religious bodies, not least the paganism of the Romans.

Ancient Greece witnessed the birth of a speculative natural philosophy based on observation and verification. The ancient Greeks worshiped the spirit of reason. Homer believed that it was the gods who urged human beings to testify to their beliefs. (Chen 2006, pp. 65–79) The adoption of this way of thinking has led to an increased reliance on reason and truth. By contrast, Christianity emphasizes pure faith and does not tolerate any act of verification that challenges the Christian religion (except the verification of God's existence by the natural light of reason). Interestingly, many Western philosophers not only understand but also support this concern of the Christians. Among them, Ludwig Wittgenstein declared openly that religious belief is not based on scientific evidence. (Clarkson 2003, pp. 52–53) Without the tradition of commitment to reason and truth originating from ancient Greece, modern science, with its spirit of rational inquiry, might not have emerged in the West. As discussed earlier, Christianity is progressive when functioning as Common Morality. However, its perspectives clash in numerous ways with the "philosophy-science" of the West. In medieval Europe under Christian rule, many Europeans scientists were punished or discouraged from pursuing their research. For example, Nicholas Copernicus, in fear of the Church's persecution, published *On the Revolutions of the Celestial Spheres*—which argues that the Earth resolves around the sun—just before his death, while Giordano Bruno, an ardent supporter of Copernican cosmology, was declared a heretic and burned in a central Roman market square. The tension between religion and philosophy-science is evident in these bloody conflicts.

What Christians are opposed to is philosophical-scientific theories that might challenge their faith, though not necessarily to technologies whose goal is to make modifications in the world to meet human needs. Despite the caricature of the Medieval ages as a supposedly "Dark Age," medieval Europe witnessed major technical advances, including greatly improved water mills and agricultural tools, the invention of windmills, mechanical printing and major technological achievement in seafaring, the textile industry, papermaking, handicrafts, as well as technological refinements with direct implications for political and economic power, such as gunpowder (Gao and Mao 2004, pp. 52–56). Friedrich Engels admits that the Medieval Ages witnessed great industrial and commercial development, together with much technological progress, including that which was imported from the East. (Engels 1971, p. 169) In less than a century, there were far more inventions developed and applied usefully than in ancient Greek times (despite that China was significantly advanced in technology and science than Europe in many ways during that era). Christianity played an essential role in regulating scientific methodology, establishing universities, and promoting the transmission of technology. European monasteries became spiritual centers which aided materially in this technological development and in building a new infrastructure which was needed for scientific communities (including centers for research, supplies, and funding) (Gao and Yaowu 2004, pp. 52–56). Meanwhile, with the support of Christians, a large number of classical scholarly texts were translated from Greek into Latin. It is said that in Toledo, the center of translation during that period, Gerard of Cremona alone, translated 87 books, including many of the works of Aristotle, Ptolemy and others. (Hao et al. 2003, pp. 65–71) All the foundations for Western modern society were laid in the medieval ages. They tremendously changed the lives of European people and paved the way for the later great scientific achievements during the Renaissance.

When Christianity gained converts on the European continent, European society as a whole tended to be culturally unified and stable. The two major threats to the life of the ancient Greeks were gradually overcome. Over-reliance on fishing was resolved through agricultural development and the rise of the manor system. And, with the material prosperity and popularity of Christianity, wars on the national level were transformed, largely moving from cultural wars to political contests that did not threaten cultural extinction. Comparatively speaking, political contests are much less destructive than cultural wars. There were once 700,000 books preserved in the Athens temples, but most of them are now lost and only a few survived the destruction of the cultural wars that led to the collapse of ancient Greek civilization. It is thought that only one-fifth of Aristotle's original works have survived. By contrast, since the Christian church survived various wars in the medieval ages, it was the church that assumed the function of protecting the classical works from being destroyed and consequently preserving the learning of Western civilization. The fact that the Germans and other tribes were converted to Christianity with the collapse of the Western Roman Empire in 476 prevented the cultural wealth in Christian churches from destruction as well. The cultural prosperity of all walks of the Renaissance was not achieved overnight, but through gradual development and accumulation that took place over the course of the medieval ages.

As it ushered in the Renaissance, the Protestant Reformation was a religious movement that had profound social impacts and a far-reaching influence upon the modern world in general. The Reformation rejected the medieval form of Christianity together with its political centralization, which was once effective in resisting the repression from the Roman Empire and promoting the transmission of religion. When the dominant position of Christianity became secure, the excess political requirements imposed by Christian churches upon believers lost validity.

Every human activity tends towards the basic principle of "maximum benefits at the lowest cost," which applies to Christianity as well. Once Christian values pervaded European society, it was reasonable for Christians to want to pay lower political cost for their religious faith. With the expanding monarchies, the political power of the church was weakened by quarrels between church and state. For example, during his reign, Henry VIII instituted the first Act of Supremacy (1536) separating the Church of England from the Roman Catholic Church. Although England still practiced core Catholic theological teachings, and retained its liturgy and the bishop system, it was under no influence from the Roman Catholic Church in the making of political policy. The necessity and validity of the church's political power was questioned and challenged once and again.

Modern Western countries impose the separation of church and state, and have a strict control on religion's impact on politics. The church/state separation is in accordance with the development of history, for they govern two different realms. On the one hand, religious liberty is maintained, since religion as "absolute faith" has authority in the realm of individual conscience, which fosters the environment where people are free to exercise their religious views and search for spiritual solace. On the other hand, the state is a political matter, an earthly reality. Only when a political system takes care of people's basic daily needs (food, shelter, medical care, etc.) can it gain firm support of its people. Therefore, politicians, unlike bishops who demand absolute obedience of his followers, would often discuss and negotiate with the citizens to ensure the maximum benefits of all social groups and bi-directional benefits between the government and the citizens. The separation of church and state is indeed inevitable, in so far as the realms that each comes to oversee (public and private) are increasingly developed.

During the long period of medieval ages, there emerged various internal efforts to reform churches. However, it was not until the eve of the Renaissance that the demand for reformation became urgent. The reformers' charges were not against the Christian faith, but against the widespread corruption in the Roman Catholic Church. (It should be noted that the leaders of the Reformation were pious Christians, and their purpose of reformation is to reform the Roman Catholic Church from within.) The translation of the *Bible* into the vernacular languages (instead of Latin) made it accessible to the common people, and ordinary Christians found evidence of the systemic corruption of the Church's Roman hierarchy. In his *Ninety-Five Theses* (1517), Martin Luther strongly protested and condemned the practice of selling indulgences, which gained support among a multitude of Christians. His efforts to reform the practices of the church launched the Protestant Reformation and thereafter, the term Protestantism was used to describe different groups that separate from the Catholic Church. Although the doctrines of these reformed churches (such as Lutheranism and Calvinism) are far from uniform, the three fundamental principles of Protestantism are unanimously observed. They are "justification by faith," the *Bible* as the only source of divinely revealed knowledge, and the priesthood of all believers. While challenging some of the fundamental doctrines of the Church as an institution, these tenets maintained and even strengthened the Christian faith, and accordingly were advocated by the common people. Many newly emergent Christian sects loosened the Church's control over individuals. Encouraged to develop a personal relationship with God—achieved by reading the *Bible*—the believers were given full control of their time, energy and wealth, which greatly promoted the prosperity of all sorts of commercial activities. The Reformation triggered a flow of new ideas that had a great impact on the modern world, evinced in prosperity in art, scholarship and economics during the Renaissance.

The Revival of the Theory of Seeking for Truth

The movements of religious reformation freed people's thinking, which made the revitalization of Greek learning possible, and in turn, the revival of Greek thinking led to the Renaissance rejection of medieval traditions coupled with a recovery of ancient philosophical tradition, especially the teachings of Plato. Socrates' commitment to knowledge and truth in Western philosophy returned in full force when the revival of Greek learning began to resurface in the philosophical thinking of Rene Descartes and Immanuel Kant.

These two philosophers stand out among others for having developed theories of the relationship between truth and religion, derived in some measure from Plato's theory of Forms (fashioned by a divine designer). Alfred North Whitehead is undoubtedly correct in claiming that "The safest general characterization of the European philosophical tradition is that it consists of a series of footnotes to Plato." (Whitehead 1979, p. 39) Descartes, who is often thought to have laid the foundation of modern Western philosophy, insists that he was discovering the "laws that God has put into nature." Although Descartes' philosophical outlook is marked and defined by the skepticism he employs in his writings, the intent of his methodological doubt is to reach the truth through the means of reason. Descartes concluded that the truth of the proposition "I think, therefore I am"—which he named "the first principle"—"is necessarily true whenever it is put forward by me or conceived in my mind" (Descartes 1971, p. 103). From this foundational thought, Descartes, with a strong motive to "know God and spirit," made a series of suppositions to argue further for the existence of God:

1. "I have in my mind the notion of a perfect being."
2. "An imperfect being, like myself, cannot think up the notion of a perfect being."
3. "Therefore the notion of a perfect being must have originated from the perfect being himself." (ibid., p. 133)

Descartes' harmonization of "the theory of seeking for truth" and "the goodness of God" (together with his ontological proof of God) is a variation on Plato's concept of the Good. Later Immanuel Kant—whose philosophical system is now regarded by scholars as "epoch-making" —declared that his goal was to "abolish knowledge, to make room for belief." Kant ended his *Critique of Practical Reason* with a famous quote that was later carved into his tombstone: "Two things fill the mind with ever new and increasing wonder and awe: the starry heavens above me and the moral law within me." "Starry heavens above" represents the objective world (the realm of "philosophy—science") and "the moral law" refers to Christian moralities, the cornerstone of Western ethical and moral system. In the case of Descartes and Kant, each of whom maintained his own Christian belief in his philosophy and life, we can arrive at the conclusion that metaphysics (as speculative science) gains harmonious co-existence with religion.

Western society cannot achieve full benefits of its science and technology without a commitment to reason and objectivity. The revival of the drive to discover the ultimate nature of reality is indispensable in the formation of Western modern science. We argue that the rise of early modern science was due to a unique combination of the commitment to empirical approaches (experimentation) and the revival of the Greek philosophical tradition. Greek philosophers attempted to seek truth. However, their claims to it were often vague—residing in the "realm of Forms" with no reference to the physical world. Later the inquiry into the nature of being (universal or ontological truth)—which was seen as the pursuit of philosophy—was dismissed in thinking during the Medieval ages. It was Francis Bacon's scientific reasoning and experimentation that provided a channel through which to understand the nature of reality by means of the critical investigation of the natural world. Under the guidance of a commitment to truth, people were free to investigate, and employ their reason to arrive at the knowledge of the laws which govern the material world, i.e., "objective truth." This sort of thinking lay at the heart of the scientific revolution, rightly bestowing on Bacon the moniker "father of modern science".

The primacy of truth combined with the method of experimentation remains the faith on which modern science and technology rests. For experimental science can only ascertain what is: it cannot justifiably speak of what should be. The theory of seeking for truth, on the other hand, deals only the question of what is a proper basis for deciding ideas and beliefs, but does not provide specific explanations of facts and relationships between facts. The coupling of the theory which seeks the truth and experimentation enables us to make scientific hypotheses based on speculation and then to validate (or falsify). There is no want of such hypotheses that have been validated or overturned in the history of modern science.

The combination of "the theory of seeking for the truth" and "experimental science" generate the modern scientific system, but not various kinds of advanced practical technology. Fortunately, after the accumulation of the Medieval ages, technological progress in the Western world provided good means for scientists to develop scientific theories, which became the basis of modern science. Among European technical advancements, the importance of the inventions of printing, gunpowder and the compass, all of which are discussed by Bacon, originated in China. Realizing that the technology of the Medieval ages made a great leap compared to that of the ancient Greek world, he wrote: "Printing, gunpowder and the compass: These three have changed the whole face and state of things throughout the world; the first in literature, the second in warfare, the third in navigation; whence have followed innumerable changes, in so much that no empire, no sect, no star seems to have exerted greater power and influence in human affairs than these mechanical discoveries." (Needham 1986, p. 9) To argue that Bacon's vision of experimental science draws its inspiration from these ancient Chinese inventions would be out of the scope of this study, however, the connection between Bacon's initiation of the idea that science can flourish through observations and through experiments and ancient China's reliance on experimental methods in technological advances should not be dismissed. In arguing for ancient Chinese contributions to scientific and technological advancement, Joseph Needham wrote: "The more you know of Chinese technology in the medieval period, the more you realize that, not only in the case of certain things very well known, such as the invention of gunpowder, the invention of paper, printing, and the magnetic compass, but in many other cases, inventions and technological discoveries were made in China which changed the course of Western civilization, and indeed that of the whole world" (Needham 1969, p. 149). Needham proves with a multitude of historical facts that ancient China had long adopted the scientific method of experimentation first advocated by Bacon in the West. Ancient China's advanced science and technology, especially the four great inventions, which opened completely new fields to the West, has had a great impact upon the development of modern technology. Without the invention of the compass, the concept of magnetism in the West might not have been formulated until much later. In spite of the general belief that pre-modern Chinese science developed without any notion of science in the Western sense, it is undeniable that China's scientific values have made considerable contributions to and still remain essential for modern science.

Conclusion

Confucianism served as a system of common moralities in traditional China, which should be given due credit. Some corruptions and decadences become rampant in China mainly as a result of the dismissal of the traditionally concrete moral values or their replacement by various other discourses in modern Chinese history. Corruption—generally associated with moral decadence—is a serious social issue which undermines the social order and stability. Although there are other means to fight corruption, the best antidote to it is moral education, for a moral agent imbued with inner moral principles ought to have the capacity to resist corruption, while a person without the bound and education in morality—even if he is kept away from corruption—would seek other means to impose damaging effects on society. Indeed, China is in urgent need of the establishment of a modern system of common moralities that is equal to Confucianism in the past.

Although the orthodoxy of Christianity was threatened in the modern period, Christian ideas and values remain at the basis of Western societies. According to Wade Clark Roof and Natalie Caron, the largest religion in the US is Christianity practiced by the majority of the population. Only 4% of the adult population identifies as having no religious belief or no religious affiliation. Roof also mentions a special group—especially in academia—who dislikes the idea of an organized religion and would very much like to claim to be "spiritual but not religious" (Clark Roof et al. 2006, p. 116).[2] This group represents the greatest increase in terms of overall population percentage—from less than 1% during the Cold War period to 10% in the beginning of the twenty-first century. They don't go to church or affiliate themselves with any religious organization: faith is private and personal, and hence religious organization is unnecessary.[3] As believers in humanism, they tend to believe in specific moral and ethical precepts of human needs. While they observe that the biblical God sometimes is too violent, they still claim that the *Bible* is the root of humanistic moralities and thus it is possible to make certain reformations of the humanism embodied in the *Bible*. Like the God in Descartes' system, the God that they worship is an embodiment of abstract conceptions such as truth, beauty and goodness deprived of its association with religion. Considering Chinese worship of a natural force called "*tian*" (which is usually translated as Heaven) in Confucianism and Taoism—as represented in this line: "(The operations of) Heaven and Earth proceed in the most admirable way, but they say nothing about themselves"—we may find commonalities between their God and Chinese "*tian*": not least in the reference to the universally common Good. With the progress of human reason, pure faith in God (this pursuit of the common Good) has tended to evolve from an image to a generalized concept.

Most ordinary people in the Western world still rely on their faith in "testaments" to practice their Common Morality. That's why Western modern philosophers such as Descartes, Kant and Wittgenstein, insist that science and religion deal with fundamentally separate aspects of human experience and should be separated so that both could serve people well. In the same vein, China needs a system of Common Moralities in the modern age while it develops modern science and technology. In fact, one important goal of the Chinese modernization is to restore and strengthen the fundamental position of morality in Chinese society. Diminishing or dismissal of traditional moral values has created a vacuum in ethical and moral standards in many aspects of daily life in China. In such a situation, some people have chosen to rely on Christianity for moral codes, so that Christianity has been steadily gaining followers over the last few decades in mainland China. We could surely learn from Christian ethics. However, as Needham has acknowledged, Christianity is unlikely to serve the development of science or the modern way of life any better than Confucianism. Since Chinese people have decided to give up Confucianism in the modern age, it does not make

[2] According to a recent survey 26% of people in Western Europe "do not believe in any higher power.".

[3] Liu Xiaofeng once said: "the so-called concept of free will means that individuals hide their affiliation with morality-religion in the 'sanctum of heart.' The inner paradox of [free but in secret] belief—such as privacy and edification, exposure and secrecy—poses a continuous problem in the history of Western thinking" (Liu 2007, p. 272).

sense for them to pick up Christianity instead.[4] What China needs is a modern system of Common Moralities that serve modern China and modern Chinese people well, which could surely learn a lot from both Confucianism and Christianity. However, it must be a new system independent of both Confucianism and Christianity, and the most important is the establishment of original moral principles and values according to the modern way of life.

Works Cited

Chen, Zhongmei. 2006. Aodesai de Renshilun Qishi [Epistemological Revelations of *ODYSSEY*]. *Waiguo Wenxue Pinglun [review of Foreign Literature]* 2: 65–79.

Clarkson, Kelly. 2003. Wuxu Zhengming, Buyong Lunzheng: Gailiangpai Zhishilun de Bianhu [Unnecessary to Testify: A Defense of Reformists of Epistemology], trans. Chen Jiaming. *Dongfang Luntan* [Forum of the East] (4).

Roof, Clark, and Wade, and Natalie Caron. 2006. Shifting Boundaries: Religion and the United States 1960-Present. In *The Cambridge Companion to Modern American Culture*, ed. Christopher Bigsby, 113–134. Cambridge: Cambridge University Press.

Descartes, Rene. 1971. *Discourse on Method and the Meditations,* trans. F. E. Sutchiffe. London: Penguin Books.

Engels, Friedrich. 1971. Ziran Bianzhengfa [*Dialectics of Nature*], trans. Yu Guangyuan. Beijing: Renmin Chubanshe.

Feng, Tianyu. 1994. *Zhongguo Wenhua Shigang* [Sketch of Introduction to History of Chinese Civilization]. Beijing: Beijing Yuyan Xueyuan Chubanshe.

Gan, Yang. 2006. *Gujin Zhongxi Zhizheng* [Tensions between Ancient and Modern as well as China and West]. Beijing: Sanlian Shudian.

Gao, Huimin, and Mao Yaowu. 2004. Lun Zhongshiji Zongjiao yu Kexuejishu de Guanxi [On Relations between Medieval Religion and Science and Technology]. *Jinyang Xuekan* [Academic Journal of Jinyang] 6: 52–56.

Hao, Liuxiang. 2003. Zhongshiji Xila Kexue de Chuanbo jiqiyu Zongjiao de Guanxi [The Spreading of Science in Greece in the Middle Ages and its Relations with Religions]. *Ziran Bianzhengfa Tongxun* [Journal of Dialectics of Nature] 3: 65–71+110–111.

Jiang, Nan. 2005. Fadong Nanbeizhanzheng de Xiaofuren: Lun Tangmushushu de Xiaowu de Lishi Diwei [The Woman who Started the Civil War: On the Historical Significance of *Uncle Tom's Cabin. Xizangdaxue Xuebao (Hanwenban)* [Journal of Tibet Univeristy (The Chinese Version)] 2: 59–61.

[4] The Christian religion now seems to be most firmly rooted in those areas to which it was transplanted to replace indigenous religion in the process of colonization: Africa and South-America. In China, the reconciliation of Buddhism and Confucianism only came after centuries of conflict and assimilation, with the result that Buddhism is adapted and unified with Chinese cultural system. The establishment of Christianity in China faces a number of difficulties: Does the establishment of Christianity have to absorb Chinese indigenous cultural elements? If so, how to adapt and assimilate Christianity in accord with Chinese traditions and Confucian ethics? If not, will establishment of Christianity on a large scale encounter resistance from Confucianism and even lead to social disturbance? The political nature of Christianity marks its distinction from Buddhism with its ascetic meditative tradition. Adaptations of Christian doctrine and worship in light of Chinese culture in history were not successful. How to construct "Chinese Christianity" to accord with the social paradigm will undoubtedly be an important task.

Li, Zhimin. 2011a. Zhengti Zhishilun: Zhishi de Zhenyude Zhibian [Knowledge-in-Wholeness: The True and the Good of Knowledge]. *Wenshizhe* [Journal of Literature, History and Philosophy] 1: 136–143.

Li, Zhimin. 2011. [Assumptions of Ontology and Development of its Research Field]. In *Zhishi de Shanyuzhen* [The Good and the True of Knowledge], ed. Li, Zhimin, 23–33. Beijing: Renmin Chubanshe.

Liu, Jiahe, and Wang Dunshu. 1994. *Shijieshi-Gudaishi Bian* [History of the World the Ancient]. Beijing: Gaodeng Jiaoyu Chubanshe.

Liu, Xiaofeng. 2007. *Rujiao yu Minzu Guojia* [Confucianism and Nation State]. Beijing: Huaxia Chubanshe.

Needham, Joseph. 1986. *Progress in Science and its Social Conditions.* Nobel Symposium 58, held at Lidingo, Sweden, 15–19 August 1983, ed. Tord Ganelines. Oxford, published for the Nobel Foundation by Pergamon.

Needham, Joseph. 1969. *The Grand Titration.* London: George Allen & Unwin Ltd.

Qian, Mu. 1994. *Zhongguo Wenhuashi Daolun* [Introduction to History of Chinese Civilization]. Beijing: Shangwu Yinshuguan.

Whitehead, Alfred North. 1979. *Process and Reality: An Essay in Cosmology.* Ed. David Ray Griffin and Donald W. Sherburne. New York: Free Press.

Xu, Songyan. 2000. Guanyu Xila Nulizhi de Lilun he Shiji [Theories and Realities of Slavery in Ancient Greek]. *Shijie Lishi* [World History] 1: 109–117.

Chapter 6
Re-examination of the Question "Where Do I Come From?": Establishment of Modern Cultural Norms Related to Parent–Child Relationships

Traditional Chinese culture emphasizes the concept of "Xiaodao (孝道)," which advocates that children be obedient to, and take good care of, their parents.[1] Where does this ethic originate? And how might it be squared with the Modern cultural ethic that promotes equality among people? Related to this, in an age where the value of equality has threatened to upend traditional, hierarchical cultures of all kinds, what obligation ought children to have to their parents? And, delving deeper, what explains the emergence and adoption of any cultural value? Addressing these questions is a must for China's smooth transformation from traditional culture to the modern. This chapter makes a contribution toward the understanding of some current and possible pathways for this transition.

As previous chapters have explored, modern Chinese culture has emerged through the convergence of several strands of traditional Chinese culture. Yet it did not emerge in isolation. As we have also sought to demonstrate, modern Chinese culture has been significantly influenced by exposure to simultaneous developments and cross currents of West culture. In order to fully grasp the relationship between modern and traditional Chinese culture, it is in fact necessary to untangle these multiple lines of influence.

The traditional Chinese cultural norm of "Xiaodao," or "filial piety" as it is often translated, and the modern cultural norm of "Equality" are both derived from different understandings and answers to a central, human question: "Where do I come from?"

[1] Xiaodao (孝道) is often translated as "filial piety" in English. However, the English translation could be misleading if one doesn't have a complete picture of Xiaodao. It is incorrect if one thinks that Xiaodao just requires children to be obedient to and take good care of their parents. In fact, Xiaodao insists that children be obedient and take good care of parents on the one hand, and it insists that parents be nice and take good care of their children on the other. The true philosophy of the culture value of Xiaodao is to balance the interest of both parents and children. Otherwise, it could not have been voluntarily observed by people over thousands of years. The philosophical principle "to balance the interest of both parents and children" is never outdated, and it should also be taken by modern culture to address parent–child relationships.

Z. Li and D. Braun, *China Being Led and Leading*,
https://doi.org/10.1007/978-981-96-6803-8_6

Answers to that question, as we will try to show, in fact structure our cultural foundation and understanding of parent–child relationships in ways that are actually more continuous than usually supposed across traditional and modern China.

Interpretation of the Origin of Humanity and Deduction of Rights and Obligations Related to Parent–Child Relationships in Traditional Chinese and Western Cultures

The most popular story from Chinese tradition about the birth of humanity is called "Nüwa creating humanity." In the version collected in *Classic of Mountains and Seas*, the story goes that "Nüwa gave birth to ten children who transformed into gods and resided in the wild of Liguang beside the road" (Zhou 2016, p. 158). In *Huainanzi*, it is said, "Huangdi gave birth to yin and yang, Shangpian gave birth to ears and eyes, Sanglin gave birth to arms and hands" (Liu 2015, p. 253). The story says that it was with the assistance of other deities that Nüwa underwent seventy attempts before finally being able to successfully create human beings. The most widely circulated version of "Nüwa creating humanity" is collected in *Taiping Yulan*, which records "It is said that when heaven and earth were opened, there were no people. Nüwa gathered yellow earth and made humans one after another with her hands. After a while, she felt short of energy. Thus she used a rope to dip into the mud, and threw the rope into air, and all the drops of mud were transformed into humans" (Li 1960, p. 365). Throughout Chinese history the story has been a touchstone of traditional wisdom: when, for instance, the problem of wealth disparity was first recognized, it was often interpreted in light of the "doctrine of fate" articulated in this story, a doctrine which explains that those born into wealth and privilege are descendants of humans created by Nüwa with her hands, and are thus nobler by birth, while descendants of those mass produced with the rope are inferior by birth and therefore destined to suffer.

Although the story of "Nüwa creating humanity" is widely circulated in the public, its influence on traditional Chinese culture was of secondary importance. Indeed, Confucian scholars, who disregarded ancient myths such as this one as forming any reliable basis for a system of cultural values and explanations, proved far more influential in forming and guiding tradition. Perhaps the most fundamental tenet of traditional Chinese values is in fact recorded in the Confucian *Classic of Filial Piety*, one of the few core texts of the philosophical school, which states, "Each inch of our body, including the hair and skin, is given by our parents, and should be cherished forever, not to be hurt in the least way. This is the beginning of Xiao" (Qu 2011, p. 1). The key phrase here is "everybody's life is bestowed by their parents," a phrase that would be interpreted as irrefutable fact, and transposed as a cornerstone of parent–child relationships in traditional Chinese culture.

Based on the premise "Each inch of our body, including the hair and skin, is given by our parents," traditional Chinese culture derived a series of cultural rights and obligations governing the parent–child relationships. The rationale of these rights can be explained as follows: (1) since the body of a child is given by his/her parents, his/her parents are the true owner of the body of their child; (2) Since the body of a child is given by his/her parents as a precious gift, the child is obliged to cherish it. In ancient times, even one's own hair was not to be cut, as it was regarded as "doing harm" to the body, and taken as violation of the spirit of "Xiao." (3) Since parents are the true owner of the body, hence the life of their child, their child has no choice but to die if the father, the master of the house, wants him/her to die. In reality, it was very rare for a father to kill a child, though it did happen in certain extreme situations. However, the absolute authority of a father over his child was regarded as unquestionable. (4) Since parents are the true owner of the body and life of their children, parents are given absolute power in deciding whom their children should marry. (5) After the death of a parent, a child is required to observe mourning for three years to show filial piety. Confucius talked about the logic behind this rule in *The Analects*: "For three years after a child is born, he/she is held in the arms of their parents. Should not a child observe mourning for three years after the death of his/her parent to show his/her deep love for his/her parents? Is not it a universal value to do so?" (Confucius 2018, p. 222) Indeed, canvassing this suite of rights, it becomes evident that the answer to the question of "Where do I come from" brings about "Xaiodao," a doctrine that would govern all aspects not only of parent–child relationships in traditional Chinese culture, but of some of the most important decisions regarding the life-direction of the child.

To promote filial piety, a popular book called the *Twenty-Four Filial Exemplars* was compiled by Confucian scholars. The work collects stories with mythological elements, such as "Filial Piety Moving the Heaven," "Lying on Ice to Catch Carp," "Tasting Excrement with a Worried Heart," each of which aims to teach people how to practice filial piety in daily life. All the stories in the *Twenty-Four Filial Exemplars* are regarded as models for people to follow in traditional China.

In traditional China, "rule by virtue" was practiced, with filial piety being regarded as the highest virtue. All virtues were upheld by both the family clan and the nation, and those who violated moral principles were punished. The patriarch of a family clan, equivalent to the father of a family, held leading power within this social unit, and the punishments for "unfilial children," under his leadership, were often harsh, resulting even in death penalty in severe cases. In cases of disputes between parents and children, the national legal system would always favor the parents. As late as in the Qing Dynasty, the principle of "the world knows no parents who are incorrect while dealing with their children" was still upheld, with the law stating that "if parents sue their children, the court should follow whatever the parents ask for without investigation" (Tian & Zheng 1999, p. 488). In other words, parents were not only favored, but presumed rightful in the eyes of the law, regardless of the facts of a case.

Similarly, the cornerstone of traditional Western culture also lies in the answering to the question: "Where do I come from?" In traditional China, the answering is from the perspective of where each person comes from, while in the West, it is from the perspective of where humanity comes from. According to the *Bible*, God created humanity on the sixth day in His own image: "So God created man in his [own] image, in the image of God created he him; male and female created he them" (Genesis 1:

27). Since God created humanity, He had absolute power over humanity, decreeing that all people should obey the will of God, which was the most fundamental principle for traditional Western culture.

The difference between traditional Chinese and Western cultures is that the latter, represented by Christianity, replaces the natural authority a parent enjoys in China with the unquestionable authority of God, and replaces the natural love between parents and children with the love between God and humanity, conceived in the form of a superordinate parent child relation. In the *Gospel of Matthew*, God explicitly states, "For I am come to set a man at variance against his father, and the daughter against her mother, and the daughter in law against her mother in law" (Matthew 10: 35). And further, "And a man's foes [shall be] they of his own household" (Matthew 10: 36). "He that loveth father or mother more than me is not worthy of me: and he that loveth son or daughter more than me is not worthy of me" (Matthew 10: 37). In traditional China, the absolute power of parents over children was always carried out in the name of love. In the same vein, the God of Christianity's demand of love is actually a demand for the utmost piety.

As traditional China had articulated powers to enforce filial piety (i.e., the authority of the father), traditional Western societies evolved powerful institutions to maintain faith in God (i.e., the authority of God and hence the church). During the lengthy Middle Ages, ecclesiastical power surpassed royal power, with religious courts possessing supreme authority, enabling them to impose severe punishments on people who did not observe religious doctrines, and even to execute those who refused to believe in God as heretics. In the ages when religion was the dominant power, all people had to obey priests or pastors appointed by the church, acting as representatives of God, in their daily life.[2]

Traditional Western culture, with Christianity as the leading faith, deduces people's cultural rights and obligations on the basis of the belief that human beings are created by God. That deduction can be summarized in the following way: (1) since humanity was created by God, all people must have absolute faith in God; (2) since humanity was created by God, all people must obey God; (3) since the church is the representative of God on earth, people must go to church regularly to listen to teachings of God, and contribute whatever is needed to the church; (4) since *the Bible* is God's book, people must believe whatever it says and follow those teachings in their daily life; (5) since priests and pastors are spokesmen of God, people must obey them.

It was in answering the question "Where Do I Come From?" that the West laid the philosophical foundation for the most important cultural principles embodied in its traditions: traditions which embodied, in their social organization, the relationship between God and humanity.[3] Indeed, the shockwaves that Darwin's theory of evolution sent through the west, ushering culture into an irrevocable phase of modernity,

[2] Russell once said, "At all times in history, the more fanatical religious belief and the more superstitious the doctrine, the more rampant cruel behaviour becomes, and the worse the situation gets" (Russell 1982, p. 24).

[3] If God, though of absolute power, had not been taken to be the creator of humanity, could it be possible for God to become the cultural faith? The answer is negative. Merely with power, a certain

were amplified by the hold that the Christian story of God's creation and stewardship of humanity exerted on the values and institutions of the West.

Exploration of the Question "Where Do I Come From?" by Modern Science: Deconstruction of Mythic or Religious Stories and Reaffirmation of Human Divinity

Modern science has explored the question of "Where Do I Come From?" with scientific methods such as observation, evidence collection, reasoning and experimental verification. Its answers to the question of "Where Do I Come From?", backed by evidence, have proven false both the mythic story of "Nüwa creating humanity" in traditional China, and the religious story of "God creating humanity" in the *Bible*.

Perhaps the most significant challenge to the story of "God creating humanity" was posed by Charles Darwin's theory of evolution, first articulated in *On the Origin of Species*, published in 1859. Because Darwin's theory of evolution is based upon compelling evidence and logical reasoning, people with reasoning power had no choice but to believe it. Although Darwin's theory of evolution itself has flaws, it provides a much more reasonable explanation of the origin of humanity than that in the *Bible*. This has greatly undermined people's faith in God as well as people's traditional cultural principles and values in the West.

In light of the theory of evolution, scientists began to probe the question of the origin of life, discovering numerous, possible answers from the reigning perspective of the day. In 1952, a graduate student named Stanley Miller, under the supervision of the Nobel laureate scientist Harold Urey conducted the famous "Miller-Urey experiment" at the University of Chicago, which demonstrated that lightning in Earth's early atmosphere could transform inorganic matter into organic matter (particularly amino acids). The experiment was among the first to provide evidence for the so-called "chemical origin of life" hypothesis. In 1969, scientists discovered 18 kinds of amino acids in a carbonaceous meteorite that fell in the town of Murchison, Australia, providing further evidence for the "panspermia" theory of the origin of life. In the late 1970s, scientists discovered several deep-sea hydrothermal vents near the Galápagos Islands in the eastern Pacific, where numerous organisms live. As a result, scientists proposed that "hydrothermal ecosystems" were ideal places for the incubation of life, known as the "hydrothermal ecosystem theory" of the origin of life. All the above theories (there are of course many others), have shed light on the possible pathways for the origin of life on Earth, and further complemented evidence of the evolutionary chain first proposed by Darwin.

In the same vein of the quests for origins, reaching back even further than life on earth, scientists have put forward theories concerning the birth of the universe. In the 1920s, astronomers discovered the phenomenon of redshift in galaxies, and

---

force may make people submit and become a political ruler, but it could never win the hearts of the people, without which one could never become a cultural faith.

thus concluded that "the universe is not static as previously assumed by many, but is actually expanding" (Hawking 2012, p. 14). As a result, the Big Bang theory was proposed, suggesting that the universe was started by a massive explosion 13.8 to 14.3 billion years ago. Currently, this theory is the most widely accepted concerning the origin of the universe.

Thus, the scientific community have put forward a set of theories concerning the origin of the universe, the origin of life, and the origin of humanity, which differ significantly from the accounts in the *Bible*. It must be acknowledged that these scientific theories are far from perfect, or conclusive. There are gaps and difficulties of explanation in every theory. For example, scientists who believe in the Big Bang theory cannot tell with evidence or confidence what the universe is like before the explosion. Similarly, scientists who put theories regarding the origin of life cannot tell which one is the pathway that life on earth had actually gone through. And Darwin's theory of evolution has weaknesses, such as a lack of transitional fossils, inconsistencies with the age of the Earth, and others. Due to the many questions unresolved, these theories, according to the definition of science itself, can only be called "scientific hypotheses," not yet "the truth" testified to by science. Nevertheless, these hypotheses are sufficient to disprove the mythic story of "God creating humanity."

With the deconstruction of the mythic stories in the *Bible* by modern science, and the declaration of the death of God in modern philosophy, the Western world experienced a period of intense anxiety. Fyodor Dostoevsky vividly expressed people's worries about life after the collapse of religious faith through the character Ivan Karamazov, who infamously reasoned, in *The Brothers Karamazov* that if God does not exist, everything is permitted. (78) However, it has been proven that, despite of the fact that mythic stories in the *Bible* have been proven to be untrue, the inherent spiritual essence of the *Bible*, which serves human society, still endures and exerts tremendous influence. In other words, although the Western world has lost much of its religious faith, especially the literal belief in the words of the *Bible*, the text still retains its cultural significance and influence, and in a sense so then does the Judeo-Christian God.

Religion was ultimately an invention of humans to serve human societies. Liu Xiaofeng stated: "The prophets of the *Old Testament* aimed to direct people's attention to human life" (Liu 2007, pp. 230–231). That is to say, the goal of prophets who tell mythic stories in the *Bible* is to redirect people's attention to live well in their everyday life. In a word, the true value of any religion is to serve people's present lives, in spite of whatever afterlife or future is posited by them. The mythic stories in the *Bible*, whether true or not, serve as a means to encourage people to adhere to social norms in daily life. For example, the miraculous stories described in the *Book of Exodus* in the *Bible*, where God personally revealed the Ten Commandments to Moses, serve as a means to strengthen people's faith in and strict adherence to these laws, which encompass fundamental social norms, such as "honoring one's parents", "not killing", and others (Li 2011, pp. 151–152). Therefore, so long as people follow social norms in their lives, a society will survive and prosper with or without the authentic religious faith that inspired those norms, and whose social function could be replaced by civil education, the rule of law, and other social institutions.

Just like religion, modern science is also an invention by humans to serve human societies. The ultimate criterion for evaluating both religion and modern science is whether they serve the interest of humanity. The goal of religion and modern science is the same, their difference lies only in the different means they employ to serve humanity. This is why some famous philosophers have proposed to avoid unnecessary attempts to put modern science and religion against each other, so as to form a kind of benign dynamic balance between modern science and religion, allowing them to benefit humanity in their own respective ways (Li 2011, pp. 135–234).

From both the religious and scientific perspectives, humanity possesses an indissoluble divinity that is of vital importance to the foundations of both traditional and modern cultural value systems. It is the divinity of humanity that makes the related cultural values divine, easy to follow, and to be kept by the public. In religion, the divinity of humanity is realized by the belief of "God creating man." In the modern age dominated by modern science, the notion of the divinity of humanity does not fade in the way that has been supposed, nor does religion exactly retreat to the background. Instead, faith in the divinity of humanity is reaffirmed, at least in the three perspectives below.

First, while modern science has grown more and more powerful, and its potential is enormous, it is forever limited by the perceptual and cognitive capabilities of human beings themselves. In other words, we get to know the world on basis of our senses. If there is something unrelated to any phenomenon we can observe, we will never be able to know it. Therefore, we have to confess that the power of modern science is finite, while mysteries of the universe are infinite. It is impossible for humanity to exhaust knowledge of the universe. In order to give a thorough answer to the question "where do I come from?", modern science must provide comprehensive explanations to all the issues concerning the origin of universe, the origin of life, and the origin of humanity, and answers which thus far remain much debated. The question "where do I come from?" has in this sense not been satisfactorily answered by modern science, and may in its very nature remain outside the scope of its powers. For the question reaches beyond human cognitive capabilities, and is imbued with infinite wonder, leaving vast space for human imagination, which grants humanity certain divinity born with them.

Modern science cannot deny that human beings could neither have been created nor survived till today without countless miracles discovered, or yet to be discovered. For example, the Big Bang theory tells that the formation of the Earth is an utter miracle. If it had not been so fortunate, the Earth might not have come into being. None other than Stephen Hawking has said, "if the hot Big Bang model has been correct since the beginning of time, then the initial state of the universe was indeed carefully chosen. It is difficult to explain why the universe should have started off in just this way unless we believe in the existence of a God who intended to create beings like us" (Hawking 2012, p. 69). Even if a modern scientist does not really believe that the miracle of the birth of Earth was brought about by the will of certain Divine Being, one could never deny the possibility in a scientific way, which, as noted above, grants humanity a certain miraculous divinity.[4] Even if one chooses not to believe the universe, the Earth and humanity were created by Divine powers, the actual miracles that brought the birth of them are sufficient to bestow a certain lofty divinity upon them. Indeed we might, as Chinese tradition has, refer to these miraculous forces as Tian (i.e., Heaven or God without a religious overtone) and recognize humans as children of Tian, to signify the divinity that these miraculous forces have bestowed upon humanity.

Second, the greatness of human life, other forms of life as well, that has been demonstrated in the process of evolution, grants life a certain divinity, even if we accept scientific theories on the origins and development of life. It is nearly incomprehensible that humanity has survived the disasters that have befallen it, and evolved accordingly. Even the most confident scientists might not be confident enough to claim that, if given another chance to repeat its evolutionary history, humanity would be so fortunate to survive and evolve into what it is today. Indeed, one could not deny that the arduous and lengthy process of human evolution and development has endowed humanity with significant divinity that we ourselves do not fully understand. Humanity might have acquired certain miraculous powers we don't really know during its evolution. To this day, modern science knows very little of the human body; the human brain, the very essence of our self-understanding, remains among the least understood dimensions of our biology, and continues to pose numerous challenges to successive regimes of modern science. We can say that human organism itself possesses a divinity worthy of our awe.

Third, the miracle of sheer survival is complemented by the miracle of sophisticated societies, without which humanity likewise would not have evolved into what it is today. In a certain sense, we can imagine each form of a human society as an Abstract Social Being, with that of all humanity as the most abstract and highest, that has harbored individual human beings. We can imagine all the hardships and disasters our ancestors experienced and all the tremendous efforts they made to have developed all the Abstract Social Beings into which we aggregate. There have been so many social disasters in the recorded history of humanity, to say nothing of those that have gone unrecorded. How have our ancestors and the Abstract Social Beings survived and developed against all of them? The long chain of inheritance that has descended through each of us is composed of countless links, strained but unbroken by the shocks of history.

Since the moment an individual is born, his or her life relies on the support of a society into which he or she is born. An individual would perish very soon without the support of a society. The society composed of and created by humanity to support itself possesses a profound divine quality that is no less significant than that of human life itself.

Besides the above three perspectives, we can further identify the divine qualities of humanity from its works, such as the arts, philosophy and more. Yet we are here

[4] Scholars in the modern age have found many errors in the *Bible* to deny the existence of God in the *Bible*. However, human beings could never prove that there is no Divine Being in the universe due to the limitation of the perceptual and cognitive capabilities of human beings.

only interested in confirming the hypothesis of human divinity; such a detailed and domain specific investigation lies beyond the scope of the present work. To clarify, the divinity of humanity helps the establishment of cultural faith which forms the cornerstone of a system of modern cultural values. Many years ago, Confucius noted: "In the past, I heard people's words and believed them. Nowadays, I hear what people say, and observe their deeds to see whether they follow their words" (Confucius 2018, p. 59). The problem Confucius encountered is the problem of a lack of sincerity and devotedness to humanity itself, one that threatens to undermine the effectiveness of cultural values, which had been well tackled by religion such as in the story of Mosaic covenant, in which Moses announced the Ten Commandments in the name of God. Indeed, religion has contributed greatly to the solidity of the Western cultural faith. For example, the key cultural faith that has been uniting the United States of America, the belief of "all men are created equal" as stated in the Declaration of Independence, takes its roots in Christianity, to be precise, in the story of "God creating humanity." It is the divinity of the cultural value, which has been granted by religion that has given it strength to help the US survive its own, numerous crises of conscience and politics.

As early as in the fourth century AD, Augustine wrote, "When knowledge that has been proven clashes with the literal interpretation of the *Bible*, the *Bible* should be interpreted metaphorically" (Barbour 2004, p. 2). This is a principle that could well balance modern science and religious faith, perhaps uniting the disparate interests and enterprises of modern societies. With this principle, modern science is given freedom to develop, so that human knowledge is constantly broadened. In the meantime, all the advantages of religion are also preserved. In fact, many people have been observing this principle in reconciling the development of modern science and religious faith. It is reported that in the US, where modern science has firmly taken root, the number of people who hold religious faith are still of great numbers (Clark Roof et al. 2006, pp. 113–134).

To people who do not have a religious tradition, like most of contemporary China, sincerity and devotedness in cultural faith could be obtained as well with the help of cultural institutions such as those in Confucianism. And the first step of course is to clarify the divine quality of humanity in the context of modern science so that people can believe it.

Reexamination of the Question of "Where Do I Come From?" and the Resulting Cultural Rights and Obligations in the Modern Age

Both the traditional Confucian answer to the question of "where do I come from" and the cultural norms that resulted from it are far from perfect. It is necessary to reexamine both of them in the modern age to produce a set of modern cultural values that bring benefits both to individual people and to societies. The perspective of a life chain previously introduced is taken up again here in a more comprehensive discussion, so that one can see the relationship of parents and children much more clearly. Thus we can more reasonably discuss the cultural values related to the rights and obligations of parents and children.

From the perspective of the life chain of humanity, we can find that parents make significant contributions in three aspects. First, parents actually give birth to children, the process through which they faithfully fulfill the hereditary obligations endowed by the life chain, contributing to the reproduction of human beings. In this sense, parents are what might be called executors of inheritance. Second, parents bring up their children by providing them with food, clothes and a place to live, which makes parents the nurturers of their children. And third, parents educate children. Most parents are able to teach their children some most basic living skills and cultural values to observe. Besides, many governments are providing good education to their citizens in the modern age. With the development of humanity, education is becoming more and more important for helping people to understand each other, bringing people together, and keeping the societies on the right track.

The contributions of parents to their children, the society and the life chain of humanity are undoubtedly of great significance. However, it is unreasonable to grant parents the cultural rights of complete control over their children in the way that the traditional cultural value of filial piety in China dictates. The reason is obvious, when viewed from the perspective of life chain of humanity: though parents are the closest link in the long chain of life to their children, they do not "own" their children as their children belong to the entire chain of life of humanity. That is, the lives of the children are the outcomes not only of their parents, but also of their grandparents, great grandparents, and so on, up to the very first origin that we could name, God or Tian. Therefore, parents do not have the cultural rights to decide their children's life and death, as it is unreasonable for them to be granted the power because their children's lives are the result of the entire chain of life headed by God or Tian, which, indeed, has made their children's lives divine, not to be placed under the will of their parents or of any other human beings.

The life chain of humanity endows each individual in the chain with certain divinity equal to each other, which concludes that children and parents enjoy equal cultural rights, in the manner propounded by the American Declaration of Independence: we are all born equal. The equal cultural rights determined by the divinity of life itself are the most fundamental in the modern age, which lays the foundation for parent–child relationships, as well as relationships among the whole population.

It is thus clear that parents, especially those in China, must recognize the rights of their children as equal to theirs, and respect them. It should be noted that the contributions parents make to their children, such as giving birth to them, feeding them, and educating them, do not grant them absolute rights over their children, if they are against the most fundamental rights of their children. Moreover, since a child does not only belong to their parents, but also to the entire life chain of humanity, human societies (not only the society in which a child is born, but all human societies) must share responsibilities in looking after the children to make sure they grow up happily and receive good education.

In traditional Confucianism, a parent's contributions to rearing children are mainly paid back at the family level, such as in the cultural institutions based upon the filial piety. However, in the modern age, these contributions have evolved to be paid at both the family and social levels. Indeed, children now often work in locales far-away from their parents, which makes it impossible for them to look after their parents as required by traditional Confucianism. Modern society, as a result, must play a part in caring for the elderly if the society wants to prosper.

In traditional Confucianism, children were asked to serve their parents at all costs, a spirit of devotion advocated in the stories "Lying on Ice to Catch Carp," "Tasting Excrement with a Worried Heart" and others. However, since children are granted cultural equality with their parents in the modern age, this service ought to be adapted to a principle that it does not undermine the divine rights granted them by Tian, a new, first principle of *modern* Xiao Dao. Therefore, children should be forbidden to carry out such acts like "lying on the ice" or "tasting excrement" in order to serve their parents, as such deeds are harming the rights of the children. In the modern age, life itself and its wellbeing are both protected by the law, which cannot be violated by a parent. And each person is given full rights in deciding whom he or she wants to marry, so that no parents are allowed to marry their child to someone against his or her own will.

It is certainly wrong for a child not to do anything in return for all the love and efforts one's parents have made to raise up him or her. So long as it does not violate the first principle of modern Xiao Dao, children are obliged both morally and legally to do something in return that would encourage future parents to invest the same, if not more, love and care they have received from their parents. In this way, the love and care that bind the rings of the life chain of humanity could be strengthened rather than be loosened. It is the second principle of modern Xiao Dao. It is a disgrace for both the child(ren) and the society if his/her/their parents who have made great contribution to their children and the society are left into a miserable situation when they are old, which will discourage people to take up the sacred hereditary responsibility that is probably the most demanding work in the world.

Traditional China had built up a cultural value system on the basis of Xiao Dao that took the traditional parent–child relationships as the core. In modern society, with the transformation of parent–child relationships described above, the whole cultural value system is to be transformed. Here we can take a look at two examples further illuminating this change.

In traditional China, the father-son relationship was taken as analogous to those of the emperor-people, thus transplanting the authority and power parents were granted over their children to that of a ruler over his people (Li 2012, pp. 61–68). Therefore, in the traditional cultural value system, just like a father who had absolute power over his children, an emperor was granted a power over the life and death of his people. However, in the modern age, when children have been granted equal cultural rights as their parents, the cultural ground for officials to claim the power of a father over their people loses its basis. Therefore, there is no alternative except for equality in the political system in the modern age. Officials nowadays could not reasonably expect filial obedience from citizens. People will never submit their modern cultural rights to someone as they were obliged to in traditional society. Equality for all has become the most fundamental faith in China which will never be shaken.

In traditional China, the parent–child relationships were taken as a model to regulate master-disciple relationships as well, a relation embodied in the common saying "one day as a teacher, lifelong as a father." In traditional China, masters were granted cultural authority to scold, beat a disciple, and in extreme cases, even to "clean up name of the house", i.e., to put a disciple to death. In the modern age when everybody is considered to be equal, the traditional master-disciple relationships have turned into teacher-student relationships that are based upon the modern cultural value of equality.

In traditional China, the parent–child relationships were taken as the most important, exclusively on the basis of Xiao Dao, which was built up to serve as the most fundamental cultural institution of traditional Chinese society. All the other traditional cultural relationships took the parent–child relationships as the cornerstone, and connected with it in a certain way to propose cultural values in the relevant fields. However, parent–child relationships are no longer so fundamental: the belief in knowledge, modern science and reason, the acknowledgement of humanity as a whole, faith in divinity of life as well as others serve as the premise for the parent–child relationships in the modern age, have all weakened this traditional cornerstone. Parent–child relationships have indeed transformed, and will continue to fundamentally transform Chinese culture and make it truly modern.

Conclusion

China has been transforming from a traditional to modern society for over one hundred years, if the May Fourth Movement of 1919 is taken as a point of departure. And the most fundamental of the transformations China has undergone lies in changes to the cultural value system that regulated parent–child relationships.

Many scholars have suggested that the best approach for China to modernize itself is simply to *copy* the West, since the West entered the modern era much earlier, and had developed into a relatively mature system that has been functional for longer than any modernity in China has existed. Unfortunately, this approach does not work well, not least with respect to the transformation of parent–child relationships. If China wants to copy the West in this respect, following a model characterized by equality, it has to rely upon Christianity and the *Bible,* which is simply not possible in China, since these foundational texts have no traditional roots in the culture. Thus China must find its own way to modernization: fortunately, there is one approach, explicated in this chapter, that could lead to the cultural faith in equality in parent–child relationships.

To build up a system of modern cultural values, the most important is a philosophy based upon reason. Fortunately, the very same Confucianism that had led traditional Chinese culture, has such a philosophy. For example, Confucius once proposed a cognitive strategy defined by the following tenet: "The most important for the development of knowledge is to acknowledge what is known and what is unknown" (Confucius 2018, p. 24). He also proposed to distance oneself from unproven stories of gods or ghosts, like that of "Nüwa creating humanity," a proposition that grants Chinese culture a certain rationality. If Confucius is an atheist at the micro level (level of individuals' actual life practice), then he is a theist at the macro level (rational cultural level). Confucius emphasized that "while presenting sacrifice to ancestors, one must do it respectfully as if the ancestors were there; and while presenting sacrifice to god, one must do it respectfully as if god was present" (Confucius 2018, p. 36). He has strong rationality in actual life practice while maintaining faith in divinity, and humility before divinity. Thus we can say Confucius is a rational theist, and his cultural philosophy is perhaps best characterized as one of rational divinity.

The philosophy of rational divinity in the modern age stresses a balance between modern science and faith in divinity, which well serves modern culture all over the world. With this philosophy, a reexamination of the Question "Where Do I Come From?", the traditional value of filial piety in China might be uprooted, and the modern cultural value of equality between parents and children re-grounded, with the reaffirmation of the divinity of humanity. This concatenation could, in turn, reconstruct obligations and rights of parents and children, as well as various related cultural values in the modern age.

Traditional Xiao Dao took roots in parent–child relationships and formed an entire kinship system on this basis. Its strength lay in the fact that it could closely unite people through these relationships, while its weakness lay in the fact that it did not posit a model of care for society outside of the kinship circle. By breaking traditional Xiao Dao, we can effectively promote people to care more of the world outside of their family circles, so that more attention, efforts and care will be paid to build up the society of China into a modern one in which all modern cultural values are well observed.

Works Cited

Barbour, Ian. 2004. *Dangkexue Yudaozongjiao* [When Science Meets Religion: Enemies, Strangers, or Partners?], trans. Su Xiangui. Beijing: Sanlian Shudian.

Roof, Clark, and Wade, and Natalie Caron. 2006. Shifting Boundaries: Religion and the United States 1960-present. In *The Cambridge Companion to Modern American Culture*, ed. Christopher Bigsby, 113–134. Cambridge: Cambridge University Press.

Confucius. 2018. *Lunyu* [The Analects of Confucius], ed. Yang Bojun and Yang Fengbin. Changsha: Yuelu Shushe.

Dostoevsky, Fyodor. 2017. *Kalamazuofu Xiongdi* [The Brothers Karamazov], trans. Geng Jizhi. Shenyang: Chunfeng Wenyi Chubanshe.

Hawking, Stephen W. 2012. *Yuzhou Jianshi* [The Theory of Everything: The Origin and Fate of the Universe], trans. Zhao Junliang. Nanjing: Yilin Chubanshe.

Li, Fang and others. ed. 1960. *Taiping Yulan* (Diyice) [Imperial Readings of the Taiping Era (Volume One)]. Beijing: Zhonghua Shuju.

Li, Zhimin. 2011. *Zhishi de Shanyuzhen* [The Good and the True of Knowledge]. Beijing: Renmin Chubanshe.

Li, Zhimin. 2012. Chongshi Lunyu: Rujiawenhuade Xiandaizhuanxing [A Reinterpretation of *the Analects*: The Modern Transformation of Confucian Culture]. *Guangzhou Daxue Xuebao (Shehui Kexueban)* [Journal of Guangzhou University (Social Science Edition)] 8: 61–68.

Liu, Xiang. 2015. *Huainanzi* [Huainanzi], trans. & annotated Hu Yajun. Nanchang: Ershiyi Shiji Chubanshe Jituan.

Liu, Xiaofeng. 2007. *Jianjin Hanzhi* [Picking All Cold Branches]. Beijing: Huaxia Chubanshe.

Qu, Xingzhi, annotated. 2011. *Xiaojing* [The Classic of Filial Piety]. Hangzhou: Zhejiang Guji Chubanshe.

Russell, Bertrand Arthur William. 1982. *Weishenme Wobushi Yigejidutu* [Why I Am Not a Christian], trans. Shen Haikang. Beijing: Shangwu Yinshuguan.

Tian, Tao. & Zheng, Qin. 1999. Daqing Lvli [Legal Cases of Qing Dynasty]. Beijing: Falv Chubanshe.

Zhou, Mingchu. ed. & annotated. 2016. *Shanhaijing* [The Classic of Mountains and Rivers]. Hangzhou: Zhejiang Wenyi Chubanshe.

Chapter 7
"Where Am I Going?": Reconfirmation of the Fundamental Cultural Faith in Life of Humanity and Its Survival

Answering the question of "Where am I going?" brings about the most fundamental cultural faith for humanity that runs through human societies from traditional to the modern ages.

Both China and the West had answered this question in complementary ways: traditionally, those answers led a future-orientated cultural faith in the life of humanity and its survival, along with a set of key cultural values which upheld it. In the modern age, a careful reexamination of the question reconfirms the fundamental cultural faith, which enables people to deal with the many unprecedented serious challenges to humanity.

Answering the Question of "Where Am I Going?" in China and the West in the Traditional Age and the Cultural Faith in the Life of Humanity

In everyday life, people can see the answer to the question "where am I going?". It is the simple fact of death. And yet there is nothing simple about this fact: it is actually one of the most complicated cultural issues, demanding careful examination.

Traditional Confucianism takes the issue of death in a pragmatic way. It does not allow emotions to overtake rationality in the face of death, and insists on handling problems brought about by death in matter of fact ways. Once a student asked Confucius: "May I ask how you think of the issue of death?" Confucius answered in an indirect way: "We even don't know much about life. How could we possibly know anything about death?" (Confucius 2018, p. 137) With this answer, Confucius implies that he is unwilling to discuss issues like death that are fundamentally unknowable, and urges his students to pay attention to that which we can in fact come to know: life. In another instance, while discussing Xiao Dao (孝道), Confucius said: "To treat one's parents well when they are alive, to bury them with proper rituals when they are

Z. Li and D. Braun, *China Being Led and Leading*,
https://doi.org/10.1007/978-981-96-6803-8_7

dead, and to serve sacrifice to them with propriety afterwards, are deeds considered to have fulfilled obligations of Xiao Dao" (Confucius 2018, p. 19). With respect to death, Confucius appears always to take a pragmatic approach.

Regarding the issue whether human beings become ghosts after death, Confucius takes a rational epistemological approach, and he confesses that he does not know it (Confucius 2018, p. 137). Similarly, while answering a question concerned with ghosts and gods, two human inventions of the afterlife, he notes: "One must fulfill one's duty to make a living above all others. As regards to ghosts and gods, we shall respect them while keeping them at a distance" (Confucius 2018, p. 79). Here Confucius makes it clear that people must not let ghosts and gods hinder them from fulfilling their duties of earning a livelihood. Confucius insists on keeping ghosts and gods at a distance, so as not to allow them to affect people's ordinary life. Indeed, Confucianism holds, in opposition to the fixation on death and its possibilities, a firm faith in life over anything else.

This pragmatic vein does not by any means preclude contemplation of death within life. Confucius is careful to allow space for mourning and sorrow for the loss of loved ones; he deems this a good way to help people to relieve the pain and recover from their losses sooner, so as to get back to the business of going on with life. Confucius himself openly expressed his sorrows in suffering the loss of close friends. After the death of his disciple Yan Yuan, Confucius felt devastated and cried aloud: "Alas! Heaven is destroying me! Heaven is destroying me!" When others suggested he had been too sorrowful, he said, "Too sorrowful? There could never be too much sorrow at the loss of this man!" (Confucius 2018, p. 136) In contrast to Confucius, Zhuangzi, one of the most important founders of the Taoist philosophy, intentionally relieved his sorrows after the death of his wife with the consolation that life and death are just two forms of the same being, like the changes of the four seasons. Different from Confucius, who holds fast to the epistemological strategy of nationality, the founders of Taoism left much space for afterlife imaginations not dissimilar to western, and especially Christian thinking.

Confucianism does not take the death of the body as the end of life. Instead, Confucius proposed there is a spiritual life, a kind of social being, beside our individual, physical lives. Confucianism holds that an individual's physical life is mortal and will inevitably decay someday, while one's spiritual life may be made immortal through self-cultivation and contribution to the greater good of society. In the view of Confucianists, for example, all those who have made contributions to society, long after they physically expired, will continue to exist in their spirit, as embodied in their words and deeds, and in this way continue to live in society and indeed influence its course. It is in this sense that they are actually alive. Confucianists believe one's spiritual life is immortal if one does good in any number of ways that are socially constructive: upholding virtues in daily life, managing actual businesses in society, or developing and expressing ideas that benefit the society. These three means toward spiritual survival are often referred to as "Three Means to Immortality:" they have been widely observed by Confucianists, and have served as motivation toward social service.

Though Confucianism contemplates the importance of physical individual and spiritual social life, it does prioritize the latter, going so far as to encourage adherents to sacrifice physical life when it comes to a choice between the physical and the spiritual. Confucius said: "Noble people would never seek to live a longer physical life at the cost of virtues. They would rather die in order not to put their names into shame" (Confucius 2018, p. 195). Indeed, there is no ambiguity here: Confucius always places the importance of spiritual life over physical life. Zengzi, a student of Confucius, ardently agreed with this view. He said: "Noble people like us must be firm in our faith, as our commitment is heavy and way to have it fulfilled is long. It is heavy because we take it as our duty to promote virtues, and it is a long way because we will observe our duty till the very end of our lives" (Confucius 2018, p. 101). Mencius, perhaps the most influential Confucian scholar, supported this view. He said: "Life is what I desire. Virtues are also what I desire. If I have to choose between the two, I would choose virtues, not life" (Mencius 2021, p. 222). The language of Mencius is clear and his mind is determined, a clarity and determination that have won him recognition among people who believe in Confucianism. Throughout the long history of China, many Confucian scholars and officials have sacrificed their physical lives to their spiritual immortality; those sacrifices have been justified by their contributions to the overall health and growth of the Abstract Social Being of China. There have been even more ordinary Confucianists who have taken famous Confucian scholars or officials as their models and have contributed greatly to this social being as well, if only in modeling it in their daily life.[1]

Confucianism has established various cultural mechanisms to sustain people's spiritual lives. The most common is the act of telling stories of those noble people who have made contribution to the society either in oral or written ways: this mechanism has not only helped ordinary people to remember those noble deeds and persons but also to plant the seeds of a noble spirit into the hearts of ordinary people. In this way, the whole society is greatly benefited. Monuments of all kinds have been established to record and highlight the deeds of those noble people as well. In certain cases, even temples have been built to worship the noble people of great deeds.

The West provided a systematic interpretation of the question "where am I going?" from the religious perspective as well. According to the *Bible*, after Adam was created, he was placed in the Garden of Eden as an immortal being. However, he, together with his wife Eve, disobeyed God. As a result, they were expelled from the Garden of Eden, and became mortal; as God said: "for dust thou [art,] and unto dust shalt thou return" (Genesis 3: 19). The story makes it clear that death is a punishment given to humanity by God for human disobedience. And it is claimed in the *Bible* that people, descendants of Adam and Eve, will be permitted to enter Heaven after death so long as one obeys the will of God.

In the *Bible*, the story of Jesus Christ's resurrection serves as an example to demonstrate that God does have the power to bring a dead person back into life. It is

---

[1] Confucianism primarily encourages noble people, in most cases, Confucian scholars and officials, to pursue an immortal spiritual life. However, ordinary people are also entitled to have immortal spiritual lives as long as they achieve certain contributions to the society.

said in the *Bible*: "…For if we be dead with [him,] we shall also live with [him:] If we suffer, we shall also reign with [him:]" (2 Timothy 2:11) Likewise: "And whosoever liveth and believeth in me shall never die" (John 11: 26). There are many other quotes that express the same faith, through which Christianity not only declares that God can overcome death but also tries all means to persuade people to believe it.

Christianity gives a definite answer to the question of "where am I going?" in the story of "the final judgment" in the *Book of Revelation.* It states that God, before the end of the world, will summon all the living and the dead before him for the final judgment, sending all good people to heaven and all bad ones to hell, destinations to which they are finally and irrevocably consigned.

All the biblical stories that answer the question of "where am I going?" really mean to urge people to treat one another with dignity first, and as a result, to contribute to a society founded on dignity. In fact, just like Confucianism, by answering the question of "where am I going?", Christianity established a socially oriented direction for individuals, an ethos that promotes social development through the cultivation of individuals.

Whether the question of "where am I going?" itself is really answered or not does not really matter, so long the fundamental cultural faith is established by the undertaking to answer this question.[2]

Modern Science's Answering to the Question of "Where Am I Going?" and the Challenges It Poses

Religion demands faith, while modern science emphasizes demonstrable proof. Modern science will not accept the premise that each person has a soul apart from the body, unless that soul can be proven to exist, i.e. to be submitted to demonstration. In the same vein, the modern scientific outlook will not brook biblical stories such as the resurrection of Jesus Christ, the final judgment, and others that are of vital importance for Christianity to uphold faith of its followers.

In terms of epistemological strategy, modern science shares the same principle of rationality with Confucianism. However, they are fundamentally different in their purposes: while the former seeks the true, the latter seeks the good. Confucianism belongs to the same category of cultural philosophy to which religion also belongs. There is another difference between modern science and Confucianism: in real life practice, both Confucianism and religions have to provide interpretations of all dimensions of human lives, including the unknown and unknowable, while modern science only needs to provide interpretations of whatever is known. Confucianism and religions are in this sense all-inclusive, while modern science devotes itself perforce to only a small part of human experience. While encountering issues of the unknown (or unknowable) like that of the soul, modern science simply assumes such things do

[2] Li argued that the most fundamental cultural faith in both traditional Chinese and Western cultures is faith in life (Li 2011, pp. 204–220).

not exist, so long as their existence is not testified in a scientific way, despite the fact that there is no scientific way to prove the existence or non-existence of such things in the first place. Confucianism does not acknowledge the existence of the soul, or ghosts or gods, in an epistemological sense. However, in contrast to modern science, Confucianism does acknowledge the possibility of their existence as well as their usefulness and influence for our lives, which is why Confucius asked his followers to "worship gods as if they were present" (Confucius 2018, p. 36). Nevertheless, unlike religions like Christianity, which personify the unknown, Confucianism keeps the unknown as unknown.

Modern scientists have carried out various scientific experiments to test the existence of souls. However, all these attempts have been failures. Many people have talked of "near-death experiences," claiming they have had experiences in which their souls left their bodies and floated in the air. In order to verify such claims, scientists have conducted research, yet no credible evidence has been found to prove that the soul exists. In 1907, Dr. Duncan MacDougall of Massachusetts, USA, designed an experiment, in which he had dying people lie on a scale and measured their weight changes. He claimed that some people's weight decreased by 21 grams at the moment of death, from which he concluded that human beings have souls which weigh roughly 21 grams. Unfortunately, the finding of Dr. Duncan MacDougall could not withstand more rigorous scientific tests, so that it was not accepted by the scientific community. In fact, no scientific experiment has so far proven the existence of the soul. As a result, modern scientists in general do not believe there is a soul, and they normally take the human body as entirely material.

On the basis of this belief, modern scientists have carried out systematic studies of the human body, obtaining greater and greater power in treating it from a biological perspective. In fact, modern science has developed a sophisticated discipline of medicine that has been so advanced that it is now posing some serious cultural challenges to society.

Modern medicine has developed all sorts of drugs and surgical procedures to treat illnesses that could have killed people in the past. As a result, the length of people's lives has been significantly increased. Modern medicine has likewise developed life-supporting systems that can keep alive a dying person for years, decades, or even longer. And further, modern science has developed techniques for freezing the body of a person, to be "resurrected" in the future, which, theoretically, can keep a person alive forever. It is obvious that modern medicine has significantly shifted the landscape of death, a landscape that was once believed to be the realm of God exclusively (in the West), or Tian (in China). The promises of modern medicine are vast, inspiring in some cases a kind of awe and reverence traditionally reserved for religious belief and practice.

As with death, the landscape of birth has also been significantly shifted in modern times. Currently, Assisted Reproductive Technology (ART), including Artificial Insemination (AI) and In Vitro Fertilization and Embryo Transfer (IVF-ET), constitute among the most radical inventions in this terrain. And medical doctors are now able to edit genes of a sperm or an egg so as to treat certain illnesses before a baby is born. The frontiers of this same technology are seemingly limitless, with some speculating as to whether gene editing can influence certain traits once thought irreducible to the biological, such as intelligence.

The nature of modern science is to seek truth, and it is powerful at discovering natural laws and designing tools to change the physical world. However, modern science does not know what is good to do or not to do, and thus it needs to be guided by a cultural philosophy in which cultural faith and values have been studied and presented and scrutinized in the modern age. For instance, modern science could have the capability to help people to live as long as they wish, or parents to produce children according to their specifications: yet only a cultural faith or value system presented by a cultural philosophy can aid us in deciding whether science (and humanity) should actually pursue these ends.

In fact, if modern medicine is allowed to help the small group of people who are in power or rich to live as long as they wish, or to help them to produce custom children, it will severely undermine the most fundamental cultural cornerstone of equality, which has guided the cultural faiths of modern societies. This might well cause great turmoil in all modern societies and cause human societies to deteriorate at a great speed. As a result, the interest of all people, including those who are in power or rich, will be greatly damaged. Even modern medicine itself that is a product of modern society will be seriously undermined. Therefore, it must not be allowed by any cultural faith or value.

It might be proposed that all people in modern societies be treated by modern medicine to live as long as they wish, and to produce custom babies. This could indeed avoid the issue of inequality. However, it will distort or even destroy the mechanism of evolution that has brought about modern human beings, which could result in the destruction of the inheritance chain of humanity, and hence humanity itself. Therefore, it must not be allowed either.

Undoubtedly, modern science has brought a lot of benefits to human beings, helping many people to live a better life. However, the problems it has brought about are serious, some of which have directly threatened the survival of humanity.

For example, cloning technology has been developing very fast. Clone monkeys have been created, and there might be scientists who are undertaking experiments to create clone human beings. A lot of literary works have expressed serious concerns about the problems cloning technology could create to the modern societies, which is horrifying, such as those in *Never Let Me Go* by Kazuo Ishiguro who has wondered about the ethics of breeding surrogates or clones who simply exist to serve human beings. Since modern science does not believe there is a soul, and since it could produce clone human beings that function the same as, if not better than, human beings, clone human beings could be justified to replace human beings, and to put human beings into a perilous situation.

Another instance of science's promise and peril is nuclear power. In the 1940s, modern science had developed nuclear weapons that had once and again pushed humanity to the very brink of extinction, and are still threating to kill humanity as a whole every minute. Furthermore, the most recent development of AI has posed another serious challenge to humanity that combines the problem of cloning with massive computer power, as some scientists believe that it will replace, enslave or simply extinguish humanity. Some scientists are now calling AI silicon-based life and human beings carbon-based life, and proposing the former, created by the latter, will take the place of the latter, just like human children, created by their parents, will take place of their parents. Some scientists believe that AI is even more dangerous than nuclear weapons in its capacity to destroy humanity as a whole, while humanity seems to be even more helpless in coping with the new threat, as very few measures have been applied to protect humanity against the threat posed by the fast development of AI.

The inner logic of modern science speeds its own development, causing it to make "progress" at ever accelerating rates. For example, after creating clone monkeys, it was a logical step of modern medicine to carry out experiments in creating clone human beings. Or, after developing ChatGPT 4, it was an obvious next step for scientists to develop ChatGPT 5, 6, 7, and so on, so long as techniques permit it. The nature of modern science drives this progress, which will constantly bring newer, and probably more deadly, threats to humanity.

It is a most demanding task to guide the development of modern science, taking its advantages while containing the threats it evolves, with the firmest faith in modern cultural values presented by cultural philosophy. It is a task the modern societies must take seriously and act fast upon.

Answering the Question of "Where Am I Going?" in the Modern Context and the Fundamental Cultural Faith in Life of Humanity and Its Survival Reconfirmed

Modern science has brought about great material wealth. However, modern science is unable to answer the question "where am I going" in a scientific way.

And yet the human view of progress requires us to hold out the possibility that there may eventually be an answer to the question "where am I going", either aided in some way by science, or other means as yet unknown to us. Therefore, in order to answer the question, the only choice we have is to keep the possibility for human beings to answer the question in the future. That is to say, by answering the question of "where am I going?", we can for now simply defer to the most fundamental cultural faith for human beings, i.e., faith in life of humanity and its survival. In other words, the conclusion we could draw in answering the question of "where am I going?" is that human beings must live on until we can answer this fundamental question.

If we suppose that human beings will never find the answer to the question of "where am I going?", would that justify the opposite conclusion: i.e. that human beings must perish as a whole along with our present rights or responsibilities? The answer is still no, because human beings could never fully prove that human beings must perish so long as there are human beings living in the world. Besides, even if human beings will perish one day, this does not deny the value of the faith in life of humanity and its survival in the present, just as the death of an individual does not deny the value of that individual's lifelong pursuit of value during his/her lifetime. In fact, faith in the life of humanity and its survival is the most fundamental value which unites West and East, traditional and modern. Indeed, so long as human beings exist, their highest interest must be their life and the survival of the species. This can serve as the most fundamental criterion for judging cultural values in the modern age, and it can be inferred that whatever helps the life of humanity and its survival is of positive value, while whatever harms it is of negative value. The dangers to humanity that are posed by those such as nuclear weapons, as well as the ideology that brought them about, are seriously threatening the life of humanity.

It is thus proven that to reexamine the question of "Where am I going?" in the modern age has reconfirmed the fundamental cultural faith in life of humanity and its survival. With this fundamental faith in mind, rules may be laid out to guide the development of modern societies accordingly.

Faith in the life of humanity and its survival is concerned with both human as individuals, and humanity as a species. The relationship between the two is dialectic. On one hand, the life of humanity as a species is made out of individual human beings. If there are no individual human beings, there is no humanity as a species either. On the other hand, all individuals will die sooner or later, while humanity as a species lives on, and ought to be maintained according to the fundamental faith articulated earlier. The value of an individual is most fundamentally decided by the service he/she provides to the society that includes him/herself.

For an individual, the cultural rights and responsibilities derived from the most fundamental faith in the life of humanity and its survival are of three major kinds. First, each individual must accept the right and responsibility "to live." So long as one is living, one fulfills one's most fundamental rights and responsibilities as an individual human being. Second, individuals must reproduce and raise their children. Only by reproduction, can humanity carry on. Otherwise, the inheritance chain of humanity will break and humanity will perish. Traditional Confucianism highly values reproduction, advocating such ideas as in the most popular idiom "the more children one reproduces, the more blessings one shall be given." It is believed that "of the three unfilial acts, not to have any descendant is the most serious" (Mencius 2021, p. 151). In traditional Christianity in the West, a child is regarded as the most precious gift from God. In the modern age, people are given more freedom to decide whether they want a child or not. However, it remains the same that humanity as a species can survive only via the means of reproduction, which must be guaranteed. Finally, individuals must work to create material and spiritual wealth. Without the former, people will be starved; without latter, a society will fall into chaos.

Longevity is a delicate issue in the modern age. In traditional societies, longevity is admired by both individuals and the societies, as the former normally have a natural inclination to live longer, and the latter, in the traditional age, could often benefit from the experiences and teachings of senior people. Moreover, the average lifespan of an individual in the traditional age was short, and even the most powerful and richest people who were taken good care of could not live much longer than the ordinary people due to the limited power of medicine. In the modern age, with the fast development of medical technologies that can help an individual to live much longer, it is necessary to look into the issue and set some cultural rules around it.

An individual's first cultural right and responsibility derived from the most fundamental faith in the life of humanity and its survival is "to live." It can thus be inferred that no one, neither the individual him/herself nor the society, has cultural right to artificially end one's life before its natural death, such as by committing suicide or by execution, i.e. capital punishment. An individual has not only rights and responsibility, in the cultural sense, to live, but also rights, in the political sense, to live a comfortable life in the modern age, as comfortable as the modern cultural values permit and one's work deserves it.

In the meanwhile, it is important to clarify that an individual has rights to have a natural death and decline any attempt to take it away. After the incident of a prolonged dying process of Ba Jin,[3] a prominent writer in China, more and more senior people have left wills to tell their children not to send them into a hospital and fill their bodies with all sorts of tubes to prolong their lives if they fall into coma and are in the process of a natural death. To further explicate why people have such rights to a natural death, we need to differentiate the cultural right from the political one. An individual's cultural rights and responsibilities are rooted in the fundamental faith in the life of humanity and its survival that end with the natural death: this can arguably be defined from the perspective of age and the integrity of the body. If an individual seeks to avoid a natural death and prolong their life, they are appealing to their political rights. Cultural rights are directly rooted in cultural faith and values, which are fundamental and must be obeyed unconditionally. However, political rights are not fundamental and can be given up by the person concerned. Therefore, senior people do have rights to insist on having a natural death.

The natural death of individuals is a natural part of the order of human societies, which is necessary for the healthy functioning of human societies. Though people mourn for the death of their loved ones, people should know that it is a natural order and it is necessary. In fact, in traditional Chinese culture, the death of a senior person is considered to be of a kind of "happy mourning," meaning that people mourn for their loved ones, yet with the belief that their beloved ones have had a happy life and are happy to leave the world, and that they themselves should feel happy too, though being sad in the meanwhile. Natural death is of positive cultural attitude and should be encouraged both culturally and politically, unless there is a special reason.

What of an individual's rights to live forever, if and when modern medicine becomes advanced enough to pose the prospect of keeping individuals young and energetic for longer and longer? If only powerful or wealthy people are allowed do this, there will surely take place a lot of social movements to overturn the system that harbors the small group of powerful and wealthy people. What will happen if even ordinary people can afford the medical expense to live forever? Societies of course

[3] Before Ba Jin passed away in 2005, he had been kept in a hospital bed, unconscious yet alive, for more than ten years, with an extremely low quality of life. It is said that Ba Jin himself claimed the right to die a natural death several times when he was conscious before becoming unconscious, which, however, was ignored unfortunately.

could not afford it, if all, or most, or many people want to live forever. This will no doubt cause a lot of social disputes in the future. Whatever decision future societies make, one basic principle should be observed: it is the order of natural evolution should be kept intact, as the survival of humanity could be otherwise endangered.

Just like in the traditional, spiritual life is highly valued in the modern age, with its importance being placed over the physical, as it is crucial for the Abstract Social Being of the modern society. A spiritual life is in fact the embodiment of cultural faiths and beliefs that guide people to build up a society. Moreover, a spiritual life, being unified with the physical life, brings about some more delicate priorities: First, positive cultural faiths and beliefs have positive impact on the body, making people happier and healthier. Second, firm cultural faith keeps people in a positive mindset, maintaining a strong vitality, and radiating a kind of spiritual charm throughout their bodies and the society. Third, cultural faiths and beliefs themselves are a certain specific information, which is expressed in the brain in a certain physical form. It is likely such information could be imprinted into genes and passed onto the next generation too.

Conclusion: Future-Orientated Cultural Faith and Values in the Modern Age

In answering the question of "Where Am I Going?", people in both China and the West in the traditional ages developed a future-orientated cultural faith in the life of humanity, which produced many cultural values, Confucian or religious, which served societies well, safe-guarding the survival of human beings and the development of social thought.

The modern age, of which the most striking difference from the traditional is the fast development of modern science, undermined some fundamental traditional cultural faiths, and posed critical challenges to the very survival of humanity - though, ironically, modern science was of course meant to bring welfare to humanity, and it indeed has brought much material wealth to humanity, if at the cost of certain spiritual deprivation. Human beings as a whole are now facing real threats of being eliminated as a species by the products of science, such as the threats of nuclear weapons, AI, pollution, global warming, clone technology, and more. All the benefits brought about by modern science of course become naught if we fail to manage the deadly threats this same endeavor has brought about.

Modern science cannot answer the question "Where Am I Going?" more scientifically in any sense than people in the traditional ages. As a result, while probing into the question of "Where Am I Going?", we have had to fall back on answers given in the traditional age, which only reconfirm the most fundamental cultural faith in life of humanity and its survival. Unlike in the traditional age, the survival of humanity is stressed in the modern age because of the fact that it is under new and greater pending threats. As a result, the fundamental cultural faith ought to draw rules to keep the development of modern science on track and to make sure of the survival of humanity in the modern age. With this fundamental cultural faith, all rules should and could be drawn to remove all the deadly threats modern science has created for humanity.

The desire and quest of an individual, a group of people, or a society for a better life for the time being have produced many political slogans and institutions. However, it has been proven that it is not sustainable if we do not pay full attention to the problems it has brought about. And it is high time for all people to observe the future orientated fundamental cultural faith in the life of humanity and its survival, and to balance the interest of individuals and their descendants, indeed, people of our generation and humanity in the future. To reach this goal, some current political slogans and institutions, as well as the rights and responsibilities of the contemporary people, need to be adjusted.

Cultural faith and values decide which are moral and which are immoral. When they become political by entering into legal terms or political rules, they begin to articulate forms of punishment for their violation. Only when real punishment for the violation of that faith can be executed can the cultural faith and values, as well as humanity in the modern age, be safe-guarded. There is still a long way to go on this path toward rulemaking and obedience, as some people have been violating cultural faith and values without being punished. This is a big problem that humanity is encountering in the modern age. Hopefully, the fundamental cultural faith in the life of humanity and its survival that benefits all political entities will be accepted by more and more people, on the grounds that all societies ought to reach political agreements that seek to safeguard the future of humanity.

Works Cited

Confucius. 2018. *Lunyu* [The Analects of Confucius], trans. and annotated, Yang Bojun and Yang Fengbin. Changsha: Yuelu Shushe.

Li, Zhimin. 2011. *Zhishi de Shanyuzhen* [The Good and the True of Knowledge]. Beijing: Renmin Chubanshe.

Mencius. 2021. *Mengzi Yizhu* [Mencius with Translation and Annotation], trans and annotated Yang Bojun and Yang Fengbin. Changsha: Yuelu Shushe.

Chapter 8
"Who Am I?": Establishment of Modern Culture on Basis of Three Forms of Personal Life

Answers to the first and second questions "Where Do I Come From?" and "Where Am I Going?" shed light on, and lead to, the third question "Who am I?" Due to the limitation of human capability in knowing the world, we are unable to give full answers to the first two questions in any complete, epistemological way. However, this does not prevent us from discussing them fully in their cultural implications, and thereby discovering the fundamental cultural faiths and values that have helped us throughout our existence to adopt answers to them: answers such as a faith in the life of humanity and its survival, a faith in the divinity of humanity, and the value of equality for all people, the latter, a faith unique to the modern age.

The third question "Who am I?" can be answered in this way: *I am a solid physical and ideological being, existing at a given moment, in three forms*, namely "the physical self," "the social self" and "the conceptual self."[1] If we call "the physical self" my first life, then "the social self" is my second life and "the conceptual self" my third one: each of which corresponds to one of the three major cultural fields, namely personal cultivation, social construction, and ideological development. It is worth considering how the promotion of modern cultural growth across these three fields or dimensions might enhance our well-being in the modern age.

"The Physical Self," "The Social Self" and "The Conceptual Self"

"The physical self" refers to the material body of a person as the result of evolution, including the brain, the torso, the limbs, and so on, as well as the needs, functions (including instincts) and other natural attributes the body possesses. The physical

[1] This chapter discusses the question of "Who Am I" from the cultural perspective. Zhimin Li discusses the same question from the philosophical perspective in another book (Li 2011, pp. 34–106).

Z. Li and D. Braun, *China Being Led and Leading*,
https://doi.org/10.1007/978-981-96-6803-8_8

self is a person's first life. It should be noted that the physical self does not include human-made articles that cover, decorate or serve as prostheses to the body, such as clothes, earrings, dentures, prosthetics, and other artificial aids. The physical self does not include any clone biological organs that might be produced and transplanted to a human body either. Once the life of a person ceases, the physical self decays and disappears from existence.

"The social self" refers to a person's social being, which we have been referring to as a kind of second life. Karl Marx proposed that a person is essentially "the sum of one's social relations" (Marx 1979, p. 5). It should be noted that, although a person's social being is embodied in various social relations, it is irreducible to those relations themselves, existing rather in and across the social totality in which those relations are embedded and develop. Each person's social being affects, which is to say alters, society – hence the notion of social totality irreducible to the individuals comprising it. That is to say, anyone has certain influence, more or less, on the society. The quality of one's social life is largely determined by the quality of the society, one's relations with those other constituent members of the society. After one's body decays and disappears, one's social being may stay and continue to influence the society as well. While one's physical self only exists in the present, the ramifications of the social self, its influence and bearing on others in reciprocal play, exist in and through time.[2]

"The conceptual self" refers to the self that exists in the ideological world. It is whatever one builds up of and for oneself on the basis of one's understanding of oneself and the surrounding society. It is a person's third life. The conceptual self includes one's self-definition, which differs, though not uninfluenced by the image imposed by society in the form of the social self. The conceptual self provides a direct guide to one's personal life, while the social image does not. The way for one's social image to influence one's actual life is in fact via one's conceptual self. The conceptual self plays a crucial role in harmonizing the relationship between "the physical self" and "the social self," helping people to find their proper positions in society. The conceptual perspective might be summed up with the equation: "I am whatever my ideas are."

"The physical self" is the very being of one's material life and its related natural attributes. It is the material base of one's existence, and of course the basis from which social life grows; as a first principle of being, so to speak, it is also the premise for the generation of one's cultural rights and obligations. The physical self is of the primary nature that serves as foundation for all, either personal or social. Hence the "first life."

"The physical self" has two major features. First, the physical self of each person is an autonomous being or closed biological system. It is evident that all individuals exist independently from the physiological perspective. Second, whatever a person feels and thinks is only *open* to oneself, and only *communicable* with others. These two

[2] The emperors in the past in China sometimes ordered the most severe punishment on a person, which includes capital punishment of the person together with all his/her family and his/her relatives, occasionally even including his/her students and friends. In this way, the emperors were killing not only the physical self of the person, but also his/her social self.

major physiological features lend credence to the notion that all human individuals are fundamentally independent and free.

The core interest of the physical self is nothing other than its existence and the full realization of its natural potential. Across traditional and the modern ages "food and sex are natural demands of human being" (Mencius 2021, p. 213). Indeed, food and sex are the two most basic needs for the existence of the physical self, with food to maintain one's material body, and sex to enable the passing of genes to the next generation. In addition, the physical self of every person has great potential to be fully realized: for example, a person is given good nutrition and physical education so that one grows tall and strong. Modern liberal parenting promotes full development of a person's mental power while traditional authoritarianism does not. The extent to which the core interest of the people's physical selves is provided for, and their potential is realized, should serve as a key indicator by which to judge the quality of a social as well as a cultural system.

"The social self" refers to one's being in a society that consists of social persons, or others. It is the unique condition of human plurality that all selves are at once in the social world both self (to themselves) and other (to another self). One does not have a social life without other people, or a society. In order to meet one's needs, one needs to seek social assistance, and to give, reciprocally, measures of that same assistance to help others to meet their needs too. The former is often granted in the name of one's rights, and the latter often required in the name of obligations. How well a society can fulfill an individual's needs, and how well it can develop an individual's potential so that he/she is able to help others, are the two most important criteria to judge the quality of the society. The social self is the extension of the existence of the physical self. In short, the social self is one's second life that one must cherish as much as one's first life.

One's social life consists of five different spheres: (1) husband and wife, a community reached via marriage; (2) the core family, a community formed by husband and wife, and their children; (3) the family group or clan, a community with blood ties; (4) the nation or state, a community defined by ethnic or political identities; (5) the human race, a community of physiological isomorphism. The quality of an individual's social life is decided by each and all of these five spheres. It is easy for an ordinary person to understand the importance of their spouses, family members, relatives and their fellow people. The other spheres are not so easily accessed or always harmonized, for historical and political reasons: i.e. divisions of race, nationality or religion and more that have made the world hard and narrow along their respective lines of division, rather than emphasize points of commonality or overlap. There are some examples to combat this tribalism: all human beings, no matter which state or nation they belong to, who are taking advantage of electricity and so are in debt to Thomas Edison, an American, who harnessed its potential. Without Thomas Edison, all people's social lives would surely be very different today. Similarly, without the four great inventions by the Chinese, the West might still be in the feudal age.[3]

[3] Joseph Needham said: "Without paper, printing, compass, and gunpowder, how could we in the West possibly move from feudal society to capitalist society?" (Needham 1969, p. 149).

The richness of a person's social life is decided by one's attention to, understanding of, and contribution to a society. The greater a person's contribution is to a society, the more valuable is one's social self, and vice versa. The social psychologist Abraham Maslow, in his theory of a "Hierarchy of Needs," proposed that the highest need of a person is self-actualization, a process that is in fact concerned with one's social life. A person social self is the most enriched, or say, a person reaches one's highest self-actualization, when one's potential is fully developed, making the greatest, as great as possible, contribution to a society, and is duly recognized.[4] It is both the right and responsibility of a person to make such contributions to a society.

The core interest of the social self lies in benign interactions and cooperations with others, and thereby with society as a whole, so as to help all individuals to achieve their highest self-actualization. Self-actualization may in this sense be characterized as a win–win, or multiple-win outcome. For society depends on the actualization and aid of all its members. How well benign interactions and cooperations are facilitated is the second key indicator according to which one ought to judge the quality of a social and cultural system.

The physical self and the social self are both related and different. They are related because there could not be a social self without a physical self. They are different in the sense that the body and its related attributes are bestowed by nature, and the growth of the body as well as the development of its related attributes follow certain rules that are relatively easy to comply with. However, the society that very much decides the quality of our social selves is man-made and historical, which is complicated and exceeds our individual control. This simple fact enjoins the creation and study of social norms and their function in the facilitation of a society's development of its individuals. The physical self ends with death. When one dies, one is no longer able to manage one's social self or to influence other's evaluations. After one dies, the society's evaluation will continuously evolve until a certain social consensus is reached. In this sense, one does not have complete control of one's social self, especially after one dies. People in power may exert more control over their social selves through the manipulation of public opinion, though that influence does not extend beyond their individual lifetimes. The much wiser way to build up one's social self is to allow people to speak and make comments freely, so that one is given a lot of opportunities to better understand others and society, in all its strengths and weaknesses. In this way, one is given opportunities to increase one's strengths and correct one's mistakes, so as to truly and effectively participate in the real building up of one's social self.

"The conceptual self" not only comprises one's understanding of one's physical self and social self, but also serves as their guide. The conceptual self grows mature after it has acquired faith in a system of cultural values that can serve as its guide. In short, the conceptual self is one's existence in the conceptual world, as the counterpart

[4] Some people accumulate a large amount of wealth by illegal means. They do not understand that the only reasonable purpose of owning wealth beyond satisfying the basic needs of the physical self is to achieve self-actualization, i.e., to gain social recognition and reputation. To use illegal means to make money directly undermines one's social self, which is obviously not desirable.

to the physical world. It is one's third life. It is by the conceptual self that one realizes one's needs and understands how to satisfy them. It is also by the conceptual self that one becomes aware of one's potential and finds the best way to fulfill that potential. Without the conceptual self, a person is a body without a soul, and is just like a puppet, easily manipulated by either their own impulses or other people's designs and needs.

To build up one's conceptual self, one needs to cultivate the habit of ardent learning, a capacity to analyze worldly issues carefully, and to think independently. One needs to learn about human nature, to understand human beings' needs, to discern good desires from bad ones, and treat them accordingly. One also needs to understand society, to be able to discern social good from evil, and firmly assist in the promotion of those goods and the fight against evil. Above all, cultivating the habit of independent thinking is the most important, because this will enable us to see through all the tricks laid to trap us, or the ill intentions to enslave us and steal our rights and due benefits from us. Without the ability to think independently, one's conceptual self can never become fully developed and mature.

To maintain the conceptual self requires one to observe the principle of being consistent in words and deeds. It is impossible for one to truly believe in one idea and act in the opposite way, because the contradictory act will automatically override the belief. A person might be able to cheat others, but one can never deceive oneself in the sense that one can always read one's own mind. If one wants to keep one's conceptual self intact, one has to avoid doing something that is contradictory to one's cultural faiths. Sometimes people might unknowingly do something inconsistent with those cultural faiths, but in this case, it is essential to repent sincerely and make atonement accordingly in order to save the faith that has been violated. Without that kind of vigilance, the faith risks losing any potency as a guide to action, and risks vanishing from the scene of individual and social engagement entirely.

The core interest of the conceptual self lies in the clarity and completeness of the system of one's cultural faiths and values. Only when a person's conceptual self becomes mature can the person's physical self and social self possibly become mature too, because the former is the guide for the latter two. To facilitate the maturation of people's conceptual selves, we must ardently promote cultural education, advocate independent thinking, freedoms of speech and of the press, and all sorts of exchanges and discussions of ideas. The percentage of people who have a conceptual self and the quality of their conceptual selves are the third key indicator to judge the quality of a social and cultural system.

The physical self, the social self, and the conceptual self are interactive, and it is important to balance them in order to promote the healthy development of all three. One is normally not supposed to promote the interest of one of the three at the cost of the other two or any other one. If one has to, one needs to consider very carefully to maximize the gain and minimize the loss.

Analyses of Problems in Traditional China from the Perspective of the Three Selves

From the ancient to the modern, the physical self of humanity and its natural attributes have changed very little. However, human beings have greatly enhanced their abilities via accumulated knowledge, scientific exploration and invention, and social reformations, thereby greatly improving the quality of life. That process has been a conscious one in terms of the self-improvements of the physical, the social, and conceptual selves.

Traditional China had both a cultural and a political system that were mature in their own ways, which could meet the basic needs of the people, and thus was a successful society in general. The most notable sign of success of a traditional society is its population growth. According to the earliest national population statistics of China available now, the population of China was 59.59 million in the Western Han Dynasty (2 AD) (Ban 1962, p. 1640). In 1850 when China was on the verge of entering the modern age, the population had reached 430 million (Ge 2008, p. 94).

However, traditional China had many problems, and plenty of room for improvement from the perspective of the modern age. In terms of the physical self in traditional China, most people's needs were met only at a low standard, and most people were not provided opportunities for the development of their potential. Indeed, as a typical agricultural society, traditional China could only provide people with materials for survival in general, though the powerful and rich, a very small portion of the population, did enjoy abundant material provision. And there were famines periodically that caused the death of many people. Besides, both the cultural and political systems were rigid in traditional China, and thus the fate of most people was predetermined from the moment of their birth: i.e. few could deviate from a predetermined life trajectory, or develop their potential in free ways. The exceptions were largely confined to the wealthy elites, and a small portion of the population able to pursue the study of traditional Confucian classics, a course that could lead to the imperial examination and civil service.

People's social selves in traditional China were underdeveloped in general. Because traditional China was an agricultural society, social mobility was very low. Most people spent their whole lives within one day's walking distance from their homes, and only ever interacted with their family and extended family members, or people living in the same village, or villages nearby. People trusted people in their respective small worlds, but they tended not to trust people or cooperate with those outside of that village life, since they had very little understanding of what lay beyond their immediate social experience and horizon. There were a lot of excellent family members for relevant families and family clans, while only a few excellent citizens for the whole society.

In terms of the conceptual self, the situation was also not optimistic in traditional China. First, only a very small portion of the population were given educational opportunities, while most people remained illiterate. Second, for the people who were educated, whatever they were taught were mainly Confucian classics, which they were required to obey, never to challenge. Third, people of original ideas were mostly either heavily censored by all the other Confucian scholars, or severely punished by the court, which took Confucianism as its ruling ideology. As a result, the general quality of the conceptual self in traditional China was low, and its contribution to the development of human knowledge was severely hindered.

There are many contributing factors to the low quality of life in traditional China, such as the underdevelopment of technology in all fields, low productivity of agriculture, and others.

From the cultural and political perspectives, there are three primary reasons. First, Confucianism in traditional China was very conservative and rigid. One key Confucian principle is subduing oneself for maintaining etiquette. Confucius emphasized this himself on numerous occasions: "To subdue oneself to maintain etiquettes is the way for reaching harmony in society. The day when all people are subduing themselves and maintaining etiquettes, the whole world will become harmonious and peaceful" (Confucius 2018, p. 146). Confucius takes self-subjugation as the prerequisite for maintaining etiquette, so that needs, especially needs at higher levels of the self, were often subdued and ignored in traditional China. Confucius's intention to keep social order by proposing measures for "maintaining etiquettes" is itself good; and the so-called "etiquettes" were really cultural norms that regulated traditional China though they were too harsh and too many. On the surface, Confucius appeared to be advocating an etiquette invented in the Zhou Dynasty.[5] In reality, he was creating a system of cultural codes on the basis of Zhou etiquettes. Zhu Xi (朱熹, 1130–1200), a great Confucian scholar, said: "The key that lies behind all the Confucian classics is simple, that is, to teach people to subdue their personal desires to keep the heavenly principles" (Zhu 2002, p. 367). The so called "heavenly principles" articulated by Zhu Xi are the same as those articulated by Confucius himself. In general, Confucianism held that personal desires are negative, and that one ought to be ready to subdue them, rather than to satisfy them. As a result, the well-being of people in traditional China was not taken good care of.

Confucianism had laid out a strict etiquette and demanded people observe it closely. One of the Confucian Classics, *The Book of Etiquettes*, recorded many such strict rules of decorum. For example, it says that children at home must "not to stay in the west-south corner of a house that belongs to the senior, not to sit in the middle of a mat, not to walk in the middle of a road, not to stand in the middle of a gate. Not to preside over meals, not to serve as the idol during sacrifices. Always to keep paying attention to the seniors and try to catch the messages even when they are not speaking a word or not giving any signs. Not to climb high, not to approach deep caves or waters. Not to laugh at people. Not to speak without making a good point" (Dai 2000, p. 2). With countless detailed rules in daily life for children and young people to observe, Confucianism in traditional China successfully cultivated obedience in them, while characteristics that are the most important in the modern age,

[5] Confucius said, "Etiquettes of Zhou dynasty were established by learning from both Xia and Shang Dynasties. How glorious etiquettes of Zhou were! I would follow them" (Confucius 2018, p. 37).

like independent thinking, free exploration of the world, and originality of thoughts, were very much subdued.

Second, the imperial system in traditional China is emperor-centered and authoritarian, which caused many deadly disasters to the society. Just like the Confucian cultural system that served the society to a certain degree, the imperial system in traditional China also played its positive roles, such as maintaining social stability and security, promulgating and enforcing various ordinances, and mitigating natural disasters, such as flood and drought. However, the imperial system had a fundamental problem: it is emperor-centered, not society-centered, and the interests of the emperor, as well as the royal family and the officialdom, were often placed over the interest of the whole society. Therefore, the society was always to some degree thwarted by its de-emphasis, and developed very slowly, though a small portion of people enjoyed power and wealth. Indeed, there were many cases in which the whole society collapsed, which caused many deaths of ordinary people, as well as a small portion of the powerful and the rich, and even emperors and their families.

The imperial system grants absolute power to the emperor, whose rule is executed through the officers charged with upholding his plan for the nation. An ideal emperor must be learned and experienced, so that he knows how to pick officials for different purposes, and how to punish corrupt officials and offer rewards to good ones; he must work hard, as there are always decisions waiting for him to make; he must be self-restrained, as he has absolute power.[6] In short, an ideal emperor must be a god who loves his people, knows everything, and works hard. Unfortunately, all emperors are just blood and flesh, and only a few of them did a fairly good job. Most of them did not live up to this standard at all. And some of them were miserably murdered, leaving the whole nation in chaos. There have been numerous shifts of dynasty in traditional China, each of which brought destruction to a large portion of the population and great social wealth.

In the imperial system, the ordinary people are supposed to be ruled, and they did not have much opportunity to develop their potential, their social capabilities or their independent thinking, so that their physical, social and conceptual selves were all constrained. In fact, the emperor and his officialdom thought that the best ordinary people could do was to work in the field to produce food to feed themselves and the rest of society. It likely never occurred to the emperors or the officials that it would be good for the ordinary people to be given opportunities to develop their potential, their social capabilities, and their independent thinking, which will in turn do good to the whole society, the emperors and the officials included.

Third, both Confucianism and the imperial system are closed and totalitarian, leaving little room for reformation and improvement either for the society or individuals. Confucianism monopolized the cultural and ideological world in traditional

[6] It is believed that the government system headed by the prime minister and the civil service system in traditional China could balance the power of the emperor to a certain degree (Zhu 1988, 1–14). However, an emperor who was not self-restrained could abolish both of them. There was never a legal document in traditional China like the Magna Carta, which limited the power of the king, signed in England in 1215. The emperors in traditional China always had absolute power; at least it was so in theory.

China. Confucianists deified Confucius, as well as some of his famous followers such as Mencius, and called them saints, which implied their infallibility. Confucian teachers demanded students memorize whatever the saints said and did, and to live a life guided by the words and deeds of these saints. They forbade students to question whatever written in the Confucian classics, and discouraged critical thinking. Confucianists, especially those who had become officials, were always ready to suppress any thoughts in the society that challenged the Confucian cultural values, ideology or life practice. As a result, generation after generation of scholars in China were tightly bound by the few Confucian classics, and almost no real breakthrough or innovation was made in the cultural and ideological field during the long history of thousands of years of traditional China. This contributed no doubt to China's so-called backwardness with respect to other nations when China encountered the West in the nineteenth century.

The imperial system did not tolerate any ideology that challenged the emperor or the officialdom, which often applied harsh means to suppress scholars who thought and spoke differently from it. The First Emperor of Qin infamously buried scholars alive who disobeyed him and burned many books he did not like, a tragedy that would be repeated by emperors to come. In the Qing dynasty, the emperor Qianlong (1711–1799) ordered a compilation of the *Complete Library in Four Sections* that was supposed to include all important books ever published. It was meant to be good, whereas Qianlong and the officials who were leading the project took this as an opportunity to destroy all books that they deemed unfavorable to the emperor and his officialdom. It was a great loss of written knowledge.

In summary, although people in traditional China were able to survive and reproduce, the standard of life was not high at all. Because Confucianism and the imperial system suppressed innovations in all fields, China made extremely slow progress during the thousands of years from the first Dynasty to the end of the last of Qing. When China encountered western powers in the nineteenth century, people in China started to realize how backward China was in comparison, and all people in China, especially the ruling class and the educated members of society, decided to catch up with the West. It was in such a situation that China decided to give up both Confucianism and the imperial system, decisions sparked by the May Fourth Movement in 1919 and the Xinhai Revolution of 1911 respectively.

Establishment of the Modern Cultural System for Welfare of the Three Selves

In comparison to those in the ages before them, both Confucianism and the imperial system in China are much more advanced. Both are a complete system that sought to guarantee social order, and both were largely successful in this respect during the ancient times in which these imperatives remained crucial. The systems, that is to say, created a relatively peaceful world in which ordinary people could work, survive and reproduce. However, both had many shortcomings, some quite serious, as discussed above. Modernity doubtless required and requires their continuous reformation.

In order to create and maintain social order, traditional Confucianism was very harsh to individuals, which is totally unnecessary in the modern age. It seemed that the Confucian sages had failed to realize that the ultimate aim for creating and maintaining social order is to serve the interests of the individuals. The key to reforming traditional Confucianism was to create a new balance between the interest of society and those of individuals. As a result, the new cultural system in the modern age needed to lose all the defining features of Confucianism such as *Filial Obedience*, *Three Obediences and Four Virtues* and others, and take such key cultural values as equality and individual independence that are typical modern cultural values, if it was to be worthy of the name of a *Modern Cultural System*.

The principle for traditional Confucianism to create order was to subdue oneself in order to maintain etiquette. The etiquette traditional Confucianism proposed was comprehensive of everyday life, and it was both rigid and trivial, so that an individual was always required to subdue oneself in order to follow it. As a result, an individual became a slave without any independent thoughts or originality. In traditional China, to subdue oneself itself became a key virtue, which resulted in many negative consequences. First, it discouraged people, especially those who were in a comparatively inferior social position, from expressing their own interests, so that good communication was suffocated. Second, it misled people to ignore personal interests, not only of the others but also of themselves, which led to the situation in which no one's interests were taken good care of in general. No means could be established for creating and maintaining any sense of social flourishing. Third, when people's personal interest was not taken good care of, they tended not to really care about the social order, or etiquette, though they might have to pretend to follow these. This is a phenomenon that can be named "moral inefficiency," that is to say, when a moral standard gets so high that people cannot really reach it, it will lose true support from the people and collapse in practice.

In fact, to subdue oneself is not necessarily a virtue. The criterion for judging whether a virtue is a virtue is to see whether it benefits the society. To subdue oneself is good when it benefits the society, while it is evil when it hurts. For example, when one hurts one's own health by subduing oneself, it can become an evil, as one is also a part of that society. Even nowadays, some people think it is virtuous not to stay at home to take a rest but to go to work when one gets ill. Such reasoning hurts individuals and in turn hurts society in the long run.

In the modern age, it is also important to create and maintain the social order. However, unlike in traditional societies, the interest of the individual people needs to be protected at the same time. A new balance between the social order and the interest of the individuals should be established, of which the leading principle is that the maximum welfare of the three selves of an individual is taken as the purpose for creating and maintaining social order, and an individual is required to subdue oneself to create and maintain social order only when it is really necessary. With this principle of regard for the care of all individuals, a society that consists of individual people can become most prosperous.

What we have discussed is the rationale for traditional Confucianism to be reformed in modern China. In reality, it has long been given up as a ruling cultural philosophy, though certain values and practices that could co-exist with the modern cultural system are still kept. In the meanwhile, the traditional imperial system that had been a close partner to traditional Confucianism was also overthrown in the twentieth century, with the modern political system assuming its place.

The key difference between the imperial system and the modern political system is between rule by man and rule of law. The greatest weakness of the imperial system was the insufficiency of the emperor to the power granted him, an insufficiency damaging to the emperors, the officialdom, and the people. To be an emperor seemed to be glorious. However, it is the most dangerous job as some of the emperors were murdered, some overworked, some depressed and subject to psychological problems. Emperors are considered to be probably one of the most unfortunate group of people in history in China. Of all the five hundred or so emperors in the history of China, only a few are considered great, and even this select group made serious mistakes and suffered profoundly (Zhang 2007). Indeed, due to the innate problem of the imperial system, no human being could really be qualified for the job of an emperor. Traditional China chose the imperial system and the emperors only because it did not have any better option. In the modern age, however, people found a much better choice of political system, which led to the abandonment of the imperial system once and for all.

In the imperial system, the officials were often suspected, punished or killed by the emperor. For example, Zhu Yuanzhang (1328–1398), after becoming emperor, infamously killed most of the generals who had followed him to overturn the Yuan Dynasty (1271–1368), just because Zhu wanted to make sure nobody could challenge either him or his heirs. In theory, an emperor could kill anybody at his will, and did not need to take any responsibility. There was a widespread saying: to keep company with an emperor is just like keeping company with a tiger, as you might lose your life at any moment. The saying reflects how helpless an official might have felt in the imperial system. Ironically, it was often encouraged for an emperor to kill some officials, with or without an excuse, so that he could rule them by terror.

However, the most miserable subjects of this system were of course the ordinary people. They had to work very hard in the field to feed themselves. Because agriculture depends heavily on the weather, they might not have a good harvest when the weather went wrong, which meant they did not have enough to feed themselves for a whole year. And when the weather went wrong for a few consecutive years, farmers starved: indeed, the history of ancient China is dotted with such mass deaths. During the warring ages, many farmers were also forced to join the army, and were killed or seriously wounded in battle. When there were major wars, like those that turned over a former dynasty (of which there were several dozen examples throughout Chinese history), a huge portion, even more than half, of the whole population could be lost.

In the modern age, the development of knowledge and technology in all fields made the modern political system possible which is characterized by the rule of law. In the modern political system, each person, including the very top leader, the officials and the ordinary people, share power together, so that it is not centered on anyone, but on the society, or to put it another way, the law itself. This has effectively overcome the greatest weakness of the imperial system.

In the modern political system, each individual's rights and responsibilities are reasonably designed and clearly given by the law, so that the very top leader of the government does not need to overwork himself. In fact, the top leader just needs to get their job done, and is free to enjoy a holiday like everybody else. Since everybody is granted certain fundamental, inalienable rights, one does not need to worry about being harmed by other people. It is certain that neither officials nor ordinary people need to be worried that their lives might be taken away by anybody. Even the most trivial legal rights of theirs are protected by the law, and cannot be taken away by anybody. At the same time, the power of each person, each official, the very top one included, as well as any institution, is strictly regulated by the law, so that no one can abuse their power to bring harm to the other people or the whole society. In the imperial system, the ordinary people and the officials are all slaves, as they don't have any power of their own. In the modern political system, everyone is the true master of him/herself as well as the society, as each of them shares some power, and is granted certain legal and inalienable rights.

Conclusion

Confucianism and the imperial system were adopted in China because they were the most competitive in ancient China. However, with the development of modern sciences in both the social and natural fields, together with the application of modern technology, much better choices have been available and have been applied in practice, which have been proven to be much better from the perspectives of both the individual people and the society as a whole.

To answer the question of "Who am I?" from a synchronic perspective, together with detailed explication of the three selves, offers a philosophical guide to how an ideal modern society should be constructed, or how a modern society should be reformed to be more ideal, so that all individual people's three selves are taken good care of. At the same time, those answers can also provide a guide to the individual concerning exactly what to do in the modern age in order to fully develop one's potential and live a life to its full extent.

With Confucianism and the imperial system as leaders, traditional China developed the most sophisticated cultural faith and value systems that regulated the nation for thousands of years. With the modern cultural and political systems in place, related cultural faiths and values have been developed to provide a guide for daily living. More must be done though. For it is only when equally sophisticated modern cultural faith and value systems have been well established and taken firm roots in the hearts of the people, that China will become a leading nation in the modern age.

Due to inertia of traditional Confucianism and the imperial system, and due to the fact that to understand both the modern cultural and political systems is pretty demanding, and that self-cultivation by modern cultural faiths and values takes great time and effort, it will take some time for any nation to become fully developed as a modern nation. It is true that the nations in the West, with the US as the leader, entered the modern age much earlier than the other nations, and have been doing well in general: yet they are far from being fully developed. For example, there are a large number of people whose three selves are underdeveloped or under cared for in the US and in the West in general. China entered into the modern age much later, and it will surely take more time for it to become fully developed, which will demand much more patience and resilience in its course to push itself forwards toward full modernity.

No matter how difficult it is for a nation to be fully developed in its modern cultural and political systems, there is no way back, as the journey to modernity for all humanity is irreversible. And yet for all that difficulty, we will be able to reach the goal as long as we continue to move forward, day after day, generation after generation.

Works Cited

Ban, Gu. 1962. *Dilizhi • Hanshu • Juanershiba* [Geography Section • History of Han Dynasty (Volume 28)]. Beijing: Zhonghua Shuju.

Confucius. 2018. *Lunyu* [The Analects of Confucius], ed. Yang Bojun and Yang Fengbin. Changsha: Yuelu Shushe.

Dai, Sheng. ed. 2000. *Liji* [Book of Rites], annotated Cui Gaowei. Shenyang: Liaoning Jiaoyu Chubanshe.

Ge, Jianxiong. 2008. Zhongguo Lidai Renkou Shuliangde Yanbianji Zengjiande Yuanyin [The Changes of Population and Their Causes in Different Dynasties in China]. *Dangde Wenxian* [Literature of Chinese Communist Party] 2: 94–95.

Li, Zhimin. 2011. *Zhishi de Shanyuzhen* [The Good and the True of Knowledge]. Beijing: Renmin Chubanshe.

Marx, Karl. 1979. *Makesi Engesi Quanji* (Disanjuan) [Marx and Engels' Complete Works (Volume 3). Beijing: Renmin Chubanshe.

Mencius. 2021. *Mengzi Yizhu* [Mencius with Translation and Annotation], trans and annotated Yang Bojun and Yang Fengbin. Changsha: Yuelu Shushe.

Needham, Joseph. 1969. *The Grand Titration*. London: George Allen & Unwin Ltd.

Zhang, Hongjie. 2007. *Zhongguo Huangdide Wuzhong Mingyun* [The Five Fates of Chinese Emperors]. Taiyuan: Shanxi Renmin Chubanshe.

Zhu, Xi. 2002. *Zhuxi Quanshu* (Dishisice) [Complete Works of Zhu Xi (Volume 14)].Shanghai: Shanghai Guji Chubanshe.

Zhu, Zongbin. 1988. Shilun Woguo Fengjian Junzhuzhuanzhi Quanlifazhande Zongqushi [A Discussion on the General Trend of the Development of Feudal Monarchy in Our Country]. *Beijing Daxue Xuebao (Zhexue Shehuikexueban)* [Journal of Peking University (Philosophy and Social Sciences)] 2: 1–14.

Chapter 9
Body Thinking, Image Thinking and Concept Thinking: The Philosophy of Thinking and Its Implication for Communications

René Descartes once put forwards the famous argument "I think, therefore I am" (Descartes 1971, p. 103). However, after a thorough discussion of what thinking is in this chapter, it is found that the philosophy of thinking itself suggests that "I am, therefore I think." Therefore, René Descartes was really saying "I am, therefore I am." That is to say, René Descartes' famous argument is in fact saying nothing.

Philosophers in China and the West tend to define thinking as a logical operation. For example, a famous dictionary of philosophy defines it this way: "Thinking is a process that the human brain reflects the real world in active, abstract and indirect ways. It includes logical thinking and image thinking. However, it often refers to logical thinking" (Fu 1984, p. 336). In this definition, logical thinking is highlighted. It seems that the editor of this dictionary, as well as many other philosophers, value logical thinking over what might be called image thinking—a form of creative and associative cogitation that does not proceed according to logical operations. Yet image thinking and logical thinking are equally important in our mental lives, though they function differently and are applied in different situations.

One of the most famous scientists in modern China, Qian Xuesen (钱学森，1911–2009) has argued that there are three different kinds of human thinking, each of which is important in its own ways: image thinking, abstract thinking, and inspirational thinking (Qian 1983, p. 564). It is recommendable for Qian to stress the importance of both image thinking and abstract thinking. However, it is not quite correct to claim that inspirational thinking is a type of thinking that is parallel to the other two. The linguist Fan Xiao (范晓) argues that "Inspirational thinking is thinking performed by the human brain in a subconscious manner" (Fan 2003, p. 76). Inspirational thinking is sometimes called intuitive thinking. On the basis of his research, A. Baylor discovered that the integration of reasoning is an important component of intuitive thinking. That is to say, he has argued that while traditionally thought of as entirely separate spheres, reasoning can be found to underlie aspects of intuitive thinking. (Baylor 1997, pp. 185–194.) Although inspirational thinking operates in a subconscious manner, it does not have any features that differentiate it essentially

Z. Li and D. Braun, *China Being Led and Leading*,
https://doi.org/10.1007/978-981-96-6803-8_9

from image thinking and abstract thinking. Therefore, it cannot be called one type of thinking that is parallel to the others.

Most scholars agree that there are at least two types of human thinking, namely, image thinking and abstract thinking (or logical thinking). Some philosophers tend to play down image thinking. René Descartes often complained that human senses are unreliable. He once said: "Since the senses have deceived us many times, I assume that none of the objects imagined with the help of the senses exist." (Descartes 2004, p. 30) It is no wonder that philosophers like René Descartes do not trust image thinking as it is closely related with human senses, drawing on the data of those senses to create even more fanciful thinking. However, there are also scholars who have emphasized the importance of image thinking. For example, Rudolf Arnheim once criticized the tradition of contempt among philosophers seeking to make knowledge claims for the human senses. (Arnheim 1986, pp. 43–44). On the basis of his own extensive work, Rudolf Arnheim discovered that "In human rational thinking, human beings are always relying on concrete images as the basis at any moment." Rudolf Arnheim proved that "all kinds of human thinking, especially original thinking, are carried out through imagery." (Li 1999, p. 69) Rudolf Arnheim's findings suggest that rational thinking (or logical thinking, or abstract thinking) is not independent of image thinking. Rather, it derives in part from image thinking. In fact, some philosophers, unlike René Descartes and his fellows, did emphasize the importance of human senses, and perhaps unsurprisingly, image thinking. For example, Aristotle once wrote: "Without the senses, we can neither learn nor understand anything. Even when we are meditating, we must be thinking of some kinds of images" (Aristotle 1992, p. 83). It is obvious that Aristotle acknowledges the fundamental role that the human senses and image thinking play in human thinking activity. Michael Polanyi went so far as to assert that all thoughts are incarnate; and that they live by the body (Polanyi 1969, p. 134). It is true that image thinking is based on human senses. So is logical thinking. Indeed, if we think of the senses as primary, furnishing data for consciousness, then it might be said that all kinds of human thinking derive crucial information if not structure itself from the human senses. It would be profitable to study all kinds of human thinking from this perspective, which does not denigrate but rather respects the contributions of the senses. Further, we risk occluding key discoveries and insight into the nature of thinking in carrying out study which ignores the role of human senses, and of the human body that produces the senses.

From an evolutionary perspective, all human activities, including human thinking, serve the fundamental purpose of human survival. This is what we need to keep in mind while analyzing human thinking.

Information Process and Body Thinking

The difference between a living being and an object without life lies in the fact that the former responds to external stimuli actively while the latter passively. The process for a living being to respond to certain external stimulus can be illustrated as follows:

External stimulus → Information process → Active response.

Human thinking is obviously included in the second stage, i.e., the information process. Therefore, human thinking can be defined initially as "an information process."

We can illustrate the information process more fully, as follows:

> Specific object is detected by certain senses... Information is formed by the senses → The nerve transfers the information to the brain → Certain image is formed in the brain with the information provided→ The brain analyses the image → The brain makes a judgment → The information of the judgment is sent to the body by the nerve → The body makes response... The relationship between the body and the specific object is changed for the benefit of the body.

This is just one cycle of human interactions with the external environment. Many cycles are activated in the course of a single, simple event. For example, when a mosquito bites us, in the first cycle, the information produced by the skin is of pain, and the image the brain forms is of a pain-image, and normally the response we make is to "look" at the specific area of our skin that feels the pain. In the second cycle, the image the brain produces is a mosquito-image, and the response we normally make is to swat it. In the third cycle, we will see whether the mosquito is killed…

Normally, thinking is considered to be of the three steps, namely, "a certain image is formed in the brain with the information provided," "the brain analyses the image," "the brain makes a judgment," especially in the last two of the three. If only images are applied during the thinking process, we might reckon that the thinking is imagistic. Image thinking is defined by a scholar as "direct thinking of objects, phenomena, or events. It is based upon senses, perception and images. It is mainly processed in the right brain" (Fan 2003, p. 76). Indeed, it has been proven by scientists that "image thinking could be carried out without any language" (Qian 1986, p. 77). In other words, the brain could think just with images.

Many animals think with images too. However, animals do not have a language. Some animals use certain codes, either visual or audio, to convey information. But they are too simple to be called a language. With language, human beings developed concepts which enable them to carry out abstract thinking.

When using language, the information process becomes much more complicated, and can be illustrated as follows:

> Specific object is detected by certain senses... Information is formed by the senses → The nerve transfers the information to the brain → Certain image is formed in the brain with the information provided→ The brain analyses the image → *The brain activates the language database* → *The brain uses language concepts for analysis* → The brain makes a judgment → The information of the judgment is sent to the body by the nerve → The body makes response... The relationship between the body and the specific object is changed for the benefit of the body.

It should be noted that all human languages are made of visual or audio symbols that are themselves physical. The linguistic symbols are just like the other external objects when they make an impact on the human senses as stimuli, and are formed into images in the brain, which suggests that abstract thinking could not occur at all without image thinking. However, unlike other objects, the images of linguistic symbols are able to activate the language database stored in the brain, which enables the process of abstract thinking. Abstract thinking is defined by some people as "a type of thinking that reasons, analyzes and makes judgments on the basis of words and the concepts expressed by them… Abstract thinking is rational, so it is also called rational thinking. Abstract thinking could not be separated from language, so it is also called language thinking." (Fan 2003, p. 77) In our daily lives, we usually apply both image thinking and abstract thinking in the course of handling even the smallest process. While the key unit of image thinking is the image, the key unit of abstract thinking (or logical thinking) is the concept, so that we take the name concept thinking as the most preferred one.

From the perspective of language, image thinking can be divided into two kinds, namely, image thinking without the involvement of language, and image thinking with the involvement of language. The former occurs in the fields of visual arts, and primarily in the media of painting and sculpture, the latter in literature. From the perspective of logic, concept thinking can be divided into logical concept thinking and non-logical concept thinking. It should be noted that concept thinking can be carried out without logic, which is testified by a lot of traditional Chinese sayings in Taoism and Confucianism. Logic, while proposed to be a predicate of human nature, originated as a concept in the West, and was adopted by China in the modern age. "Logic is commonly referred to the logic in abstract thinking, i.e., the logic that governs human rational thinking (it can thus be called rational logic). It is a method that helps people to analyze, reason, synthesize, make judgements and so on. The basic unit for logic is the concept… The main types of rational logic include formal logic, mathematical logic, dialectic logic" (Fan 2003, p. 79). Wang Lu emphasizes that "the inherent mechanism of logic is… 'a conclusion inevitably to be drawn'." (Wang 1999, p. 51) In other words, logic is the principle that allows a correct conclusion to be reached from correct premises. Western philosophy has traditionally taken *truth* as its ultimate goal, a goal which can only be reached through the operations of logic, with which it is bound up. This is why Western philosophy has traditionally emphasized the importance of logic, as well as rationality.

Both image thinking and concept thinking are essentially an information process that results in judgement-making, with or without bodily response. Do human beings have any other ways of thinking that process information in related ways?

Indeed, human beings do have another way to process information: a process we typically call "instinct." The most prominent feature of instinctual thinking is that it does not go through the brain. For example, we instinctively blink when foreign objects enter our eyes. In this case, we make active responses to external stimulus without the participation of the brain. The information process of such cases can be illustrated as follows:

> Specific object is detected by certain senses... Information is formed by the senses → *The region of the body concerned analyses the information* → *The region of the body concerned makes a judgment* → The body makes response... The relationship between the body and the specific object is changed for the benefit of the body.

This information process performs functions as image thinking and concept thinking, though with different defining features, meriting the name therefore of a different kind of thinking.

For people who hold the stereotypical belief that thinking is the privilege of the brain, it may be difficult to accept the view that an instinctive response is a kind of thinking. However, the viewpoint that takes thinking as a privilege of the brain is only a hypothesis without any solid scientific basis. Since an instinctive response shares the same essence as image thinking and concept thinking, it makes good sense for us to give up the unprovable hypothesis and take an instinctive response as a kind of thinking. In fact, if we agree that human beings have undergone an evolutionary process from lower to higher and more complex lifeforms, we have to agree that human beings were once in a stage when we did not have an organ that could be called brain, just like certain lower lifeforms we can observe today. Only with acknowledgement of instinct as a kind of thinking, can we have a complete chain of different types of thinking, namely, instinctive, image, and concept thinking. When human beings were in the evolution stage when we did not have a brain, our survival had to depend totally on instinctive thinking.

It may be profitable to think of bodily thinking, instead of instinctive thinking, as the name for the kind of thinking that helps human beings to actively respond to external stimuli without the participation of the brain, because it includes other ways to process information besides the instinctive response. Rudolf Arnheim argued that human beings have thinking abilities that could be extremely powerful occasionally at the sensory level, similar to that in the brain. He wrote, "Perceptual thinking at the sensory level can also achieve what is called 'understanding' in the field of rational thinking. Anyone's eye strength can display the admirable ability of artists in a simplistic way, which is the ability to create diagrams that can effectively explain experiences through organization. Therefore, eye strength is also comprehension ability" (Arnheim 1984, p. 56). If we observe our daily experiences closely, we can find that body thinking is actually assisting us constantly. For example, each of our eyeballs jumps around a hundred thousand times a day on average, but the visual images we *actually see* each day are very limited. This means that the eyeball performs a kind of body thinking, i.e., to process information and make judgment by itself, so that it filters all the information it receives and only selects and sends more important information to the brain *to be actually seen.* Another example is that our skin constantly receives stimuli from moisture, particles in the air, and other material inputs. However, it only selects and sends certain stimuli to the brain for contemplation. It must be acknowledged that "choose not to send to the brain" is a critical judgement made by the senses that helps the brain to work much more efficiently. If human beings did not think with our bodies, our brains might be constantly over-worked or even stop working just like a virus-infected computer. Such daily phenomena, encapsulated in phrases like listening but not hearing, looking but not seeing, might all be said to be outcomes of the body thinking, which are important for our daily life. In phenomena of sensory adaptation, like being insensitive to a fragrance after staying in an orchid room for a long time, body thinking demonstrates a high-level thinking ability to actively adjust indicators for its judgement-making process, so that it does not send unimportant information to the brain.

Only upon acknowledging body thinking can we hope to develop a complete picture of how different kinds of thinking modes cooperate with each other in our lives, which might be sketched out as follows: (1) body thinking carries out a preliminary analysis and judgment on sensory information formed by external stimuli, and makes a non-action response to a large amount of unimportant information. In the meantime, it selects and sends the important information to the brain. When it detects a dangerous external stimulus, it responds instinctively. (2) Image thinking analyzes and makes judgements on some information, and makes active responses accordingly. (3) For some information that image thinking cannot fully deal with, concept thinking makes further analysis, so that we can understand it more profoundly, and make more accurate responses. It is thus clear that human thinking not only takes place in the brain, but also in the whole body. Human beings have to rely on the cooperation of all kinds of thinking in order to survive and enjoy their lives.

To clarify the nature of human thinking, and to acknowledge body thinking as parallel to image thinking and concept thinking as having significant implications, more work that is beyond the scope of this chapter must ultimately be done. Here, we only seek to make a preliminary case for this acknowledgement.

Image-Encoding and Rules of Thinking

The three kinds of human thinking vary in the methods with which they are applied. However, they share the same essence as an information process.

Children might think there is something like the actual objects in our brain when we see them. However, adults understand the difference between the world in and outside of our minds. Whatever is processed in our brain is just the information provided by our sensors. This is why we can still *see* an object in our mind even when we close our eyes: it is the information processed in our brain that makes us *see* it. For the same reason, we can recall and see an object that we saw a long time ago. It is just because our brain activates the information related and stored in our brain. Sometimes we can only recall and see part of the object, which is because our brain has lost the information of the other parts. We can also combine information taken from different objects to create an object in the brain that does not exist in reality, such as a unicorn in the West or a Loong[1] in China. This is the philosophy of thinking that explains how artists create novel images, either visual or audio in their brain.

[1] Loong (龙) is sometimes translated to be dragon, which is totally wrong, as they are just two different beings. A typical Loong is a god of rain who is supposed to help farmers to have plenty yet not too much rain so that the farmers could harvest. As China was an agricultural nation, Loong was very important and was thus often worshipped. Though there were evil Loong, a typical Loong is good and worshipped. On the contrary, dragons are almost all evil in the West.

Since all images in our brain are just information gathered and formed by our senses, they can vary from person to person, as different people have different sensory capabilities. For example, people with normal eyes and people who suffer from color blindness have very different images of the same object in their brains. Even the same person can get very different images of an object when he/she is in different moods, or ages, or impacted by other conditions that affect the information his/her sensory organs produce and the nerves that transmit it. For the same reason, the images of the same object vary enormously between human beings and other animals, as they have completely different sensory organs and nerves.

In our daily life, the mechanism for us to *identify* an object is actually for us to compare the image we get from the object with images stored in our brain. For example, when we see a rose, we get its image and compare it with all the related images stored in our brain. We identify it as a rose when we find its image is identical with the images stored in our brain that we name rose. Some senior people might lose their storage or recall of images in their brains, which will result in the fact that they cannot identify objects, from roses to their family members. When a person meets something for the first time, whatever he/she does is not identifying it. Instead, whatever he/she does is to store the image perceived in his/her brain, and with a name provisionally attached to it. When it reoccurs in the future, the person will be able to identify it instantly. Children are always full of curiosity about the world, and like to look, smell, and touch all the novel objects they meet, if possible. This is crucial for the growth of the children, as they are storing the images concerned in their brain, which will enable them to later identify them when they reoccur. Children who have a larger storage of images are surely better positioned to create more interesting artistic works.[2]

When children store an image in their brains, they label that image with a corresponding name, which becomes the foundation of image thinking and concept thinking. For example, when a child actually sees a rose and is told by his/her parents that the object is named "rose," he/she stores the image with a label "rose." Next time, when he/she reads the word "rose" or hears someone mentions "rose," his/her brain will pick up the rose he/she has actually seen in real life to make him/her *see* the image.

When human beings create a language, one most basic step is to encode all images stored in their brains, i.e., to give a name to each image. This seems to be simple. However it is very complicated. Human beings have applied various methods to encode images, such as encoding images according to relevant objects, and then according to the nature of the objects concerned, and further encoding encoded images for all different purposes. The French experimental psychologist Jean Piaget argued that cognition is "a continuous construction," (Piaget 1981, pp. 19–20) which may well describe the encoding process referred to here: it is an always continuous,

[2] In China, many parents do not understand the importance for children to be exposed to the outside world. They might over protect their children, or demand their children to stay at home to read books, both of which cause their children not well exposed to the world, and not well developed intellectually.

never completed process, coterminous with individual and human species discoveries and developments in knowledge.

The world of the language of human beings is not identical with the real world. On the one hand, the real world is infinite while the world of any language is finite. On the other hand, there are images and stories in the world of human languages that do not exist in the real world.

Human languages have their own features that are different from that of the real world. For example, Ferdinand de Saussure argued that concepts are purely representational differences. A concept cannot be defined directly according to its content. It can only be defined indirectly according to its relationship with all other concepts in a system. Its most assured defining feature is that it is not something else (Saussure 1980, p. 163). In other words, to define A, the best way is to tell it is not "B, C, D, E…Z" in the same linguistic system. The universe is infinite, while the world of language is finite, which allows us to define a concept in this way. As concepts are the key unit of concept thinking, it is crucial to have clear definitions for concepts in order to think well in this mode.

No language is established all at once. Rather, language evolves through a process of encoding and re-encoding, and, after many years, finally become a construct that works well, yet will continue to develop alongside with the evolution of the society which speaks it.

The images acquired by an individual depend on one's life experiences, and are thus personal. However, all languages are social beings. People often encounter two situations: (1) one stores an image in one's brain without knowing its name (we call such an image "name-pending image"); (2) One learns a name via a language, which one has not yet met it in real life (we call such a name "image-pending name"). An individual's learning experience is thus a process of "matching images and names" to a certain degree, which paves the way for the employment of image thinking. Likewise, we also learn to match concepts and their definitions for the development of concept thinking.

All thinking rules are implied in this methodology to encode images and define concepts. For example, the law of identity is basic in formal logic, which is made possible by the principle of one-to-one correspondence when we are encoding images or defining concepts. For another example, Aristotle's syllogism works only when the principle of property is observed when encoding images or defining concepts. This principle is demonstrated in the classification as: if a_1, a_2, a_3, a_4, a_5, a_6 … have a common property (x), then they can be collectively referred to as A. Only on this basis does Aristotle's syllogism stand, that is: if A has property (x), and a_n is included in A, then a_n also has property (x). Rather than saying that the syllogism is a rule of thinking, it would be more accurate to say that the principle of property must be followed when encoding images or defining concepts, which makes syllogism work.

If we did not have definite names for images or clear definitions for concepts, we would never think clearly. Many concepts in traditional Chinese culture, such as *Tao* (道), *Qi* (气), *Gu* (骨) among others, were not clearly defined, so that it was difficult for people to build up any solid theoretical framework. In the modern age, it is necessary to redefine all traditional concepts in order to inherit them, modernize them and carry them on.

As each concept is defined by its relationship with all other concepts in a certain language, when one culture imports a concept from another culture, it is always important to redefine it, so as to help it fit into the target cultural system, making the latter at the same time more inclusive.

Information Connection and Knowledge

The information processes of all human thinking, including body thinking, image thinking, and concept thinking, result in certain information connections which underlie human cognition and knowledge. The information connection in body thinking is inherited, while those in image thinking and concept thinking are largely acquired or learnt in real life. For example, the most common connections among images in our brain are based upon the relationships of the relevant objects in real life, such as those among leaves, branches, trunks, and roots of a tree. We call this kind of information connection, which is derived from objects in real life, natural connection that results in a kind of natural (image) thinking.[3] This kind of connection is acquired via our observation of the real world, which could be illustrated as below:

> Relations between objects in real life → Relations between images in brain → Information connection → Knowledge.

If the information connection is learnt from reading, such a process can be illustrated as:

> Linguistic signs → Images activated → Relations described in the language → Relations between the images in brain are established accordingly → Information connection → Knowledge.

We call this kind of information connection linguistic connection, which results in a kind of linguistic (image) thinking. We can find a lot of examples from literary works. In traditional Chinese culture, there is abundant knowledge that is of the natural (image) thinking and linguistic (image) thinking.

For image thinking, the basic unit for information connection is the image. However, for concept thinking, the basic unit is the concept that is formed on the basis of images. An image in the brain carries the all-round information contained in the image in real life, so that image thinking becomes a kind of panoramic thinking, the advantage of which is that it can carry multiple meanings. The disadvantage is that this kind of thinking is sometimes vague and can be misleading. Different from images, a good concept is always precisely defined so that it is itself and only itself. Therefore concept thinking in principle is always clear, and does not allow ambiguity. In order to safeguard concept thinking, a whole discipline has been well developed

[3] Cognitive linguists have invented a theory of schema on basis of such human thinking methodology.

that is called logic, which has laid out all sorts of laws for people to observe during the process of concept thinking. Concept thinking that is based upon concepts and regulated by logic is rational thinking, also known as logical-rational thinking that highlights that the process must follow the laws of logic. If we call image thinking best for aesthetic enjoyment, logical-rational thinking is best for seeking truth.

Image thinking and concept thinking have different features, and are good for different purposes, as suggested above. Some philosophers, represented by Plato, who looked down upon poets and poetry, value concept thinking and play down image thinking, which impoverishes the range of human capacity. Albert Einstein, an eminently logical thinker, wrote: "Imagination is more important than knowledge, because knowledge is limited, while imagination encompasses everything in the world, drives progress, and is the source of knowledge evolution. Strictly speaking, imagination is a real factor in scientific research" (Einstein 1976, p. 284). As image thinking is a kind of panoramic thinking, imagination encompasses everything in the world. If we compare image thinking to the sea, concept thinking is like a river. The former is rich in information while the latter is advantageous in terms of the reliable, straight and forking route it provides. Qian Xuesen also recognized the value of image thinking. He wrote: "image thinking is broader than abstract (logical) thinking. Logical thinking could solve scientific problems, while image thinking could make good use of knowledge before their being made to be scientific" (Qian 1986, p. 21). Science is limited, while the world is infinite. Concept thinking can only work well in fields that highlight rationality, while image thinking is applicable to the entire world of life. It is no good to praise one and play down another, though. Each has its own advantages and can complement the other, and both should be applied together to promote the growth of human knowledge.

From the perspective of the philosophy of thinking, human knowledge is a kind of information connection. In the field of natural science, scientists attempt to make sure that information connections are identical with those in reality, so that they often carry out experiments to verify whether a certain information connection conforms to a certain objective relationship. Such objective relationships can be testified to be true or false, so that the corresponding information connection can be called true–false connection. Friedrich Schlick once said: "A proposition can be testified whether to be significant or not only under the following conditions: it can be testified whether it is true or false with an experiment… Otherwise it is hollow and meaningless." (Hong 1964, p. 268) The judgment is correct, yet only relevant to the field of natural science, as the information connections in it are those which are true or false. In the field of arts and humanities, and social sciences, knowledge is not a true–false connection. Rather it is of good-evil or beauty-ugliness connection and so on. For example, in the proposition "all lives should be respected," the true–false criterion does not apply. It is a modern cultural faith which is appropriate to the modern cultural context. In the field of natural science, people seek truth, while in the field of arts and humanities and social sciences, people seek meaning.

The most precious of human thinking lies in the innovation of making new information collections. In image thinking, innovation occurs when new information connections of images are made to produce new images. In concept thinking, to innovate, i.e., to make new information connections, often results in the production of new concepts and new links between concepts. In many cases, we need to use both image thinking and concept thinking to innovate. For instance, we apply image thinking to imagine a straight line that we cannot find in reality, as there is no absolutely straight line outside the imaginary realm of mathematics. We apply concept thinking to conceptualize a straight line and give it an accurate definition. As a result, a key concept is innovated, which is of vital importance in mathematics and other related fields. It is with all sorts of innovative thinking activities that human beings have created a world of ideas and have greatly enhanced the real world with all sorts of inventions.

In order to help us think better and produce more knowledge, there are many steps we must continue to take. First, it is desirable for us to have proper physical exercises so that we could have good senses and nerves, with which the body thinking could work well. Second, it is desirable for us to be exposed to all different sorts of environments so that we can experience abundant images and store them in our brains. The more images we have stored in our brains, the more resources we have for future innovations. Image thinking is the basis of concept thinking, which means one cannot have good conceptual thinking without good image thinking. Third, to become a good thinker of concepts, one needs to carefully grasp all different concepts and their connections, which form a world of ideas of themselves. The mission of the scholar is to make the world of ideas as perfect as possible, while it is the mission of the politician to make the real world as good as possible by following the model of the world of ideas.

Conclusion

Thinking has been regarded by most philosophers, such as René Descartes, as capacity specific to the brain, which, nevertheless, is untrue. In fact, the nature of thinking is an information-process, which reaches far beyond the brain, highlighted by judgment-making within the human body, of which the other parts of the body, besides the brain, are also capable. The capacity of body thinking suggests that human beings are thinking so long as a human being are alive, which suggests that "I am, therefore I think." If this is the case, René Descartes's famous argument "I think, therefore I am" is of no significance at all.

Whatever grants human beings thinking capacity must lie in the cognitive construct passed from one generation to another by means of material genetic inheritance, without which, (as with non-human animals), we could not develop human-like thinking capacities. Apart from the cognitive construct, humans have also inherited some simple yet crucial concrete processes, namely instinctive knowledge (such as those demonstrated in actions of unconditional reflex), so that a new life can be started, which is shared by many animals other than human beings. With innate thinking capacities as a foundation, human individuals can enhance their thinking ability to a very sophisticated level by learning and refining it during their lifetime.

Human thinking can be generally divided into three distinct, yet interrelated categories, i.e., body thinking, image thinking and concept thinking, all of which serve the fundamental purpose of human survival. In accordance with the evolution of human beings from basic to more advanced, our thinking capability has also undergone a similar procedure, namely, that of body thinking to image thinking, and finally to concept thinking. The human brain came into its sophisticated, linguistic state of being in a relatively late period of its own evolutionary history, when the cells gradually developed a specific organ to deal with complicated information, and with more efficiency, which became the basis for the development of image and concept thinking.

Human beings would most trust, instinctively, body thinking over the image thinking, and over concept thinking that is characterized by logic and reason. Concept thinking, which has always been highly emphasized at the cost of the other two, in fact, is far from being perfect and less significant than the other two in terms of serving the survival of human beings. Philosophers, scientists and critics who take reason, something begotten by the concept thinking, as the highest principle, have produced many theories of literature, or art at large, which, however, is essentially based on the body thinking and the image thinking. Accepted concepts granted in many theories of literary studies or cultural communication that are more or less distorted by the mere fact of emphasis of concept thinking over the other two, can be well modified or corrected in light of these propositions.

Works Cited

Aristotle. 1992. *Yalishiduode Quanji* (Disanjuan) [Complete Works of Aristotle] (Volume 3), trans. Miao Litian and others. Beijing: Zhongguo Renmin Daxue Chubanshe.

Arnheim, Rudolf. 1984. *Yishuyu Shizhijue* [Arts and Visual Perception], trans. Teng Shouyao. Beijing: Zhongguo Shehuikexue Chubanshe.

Arnheim, Rudolf. 1986. *Shijue Siwei* [Visual Thinking], trans. Teng Shouyao. Beijing: Guangming Ribao Chubanshe.

Baylor, A. 1997. A Three Component Conception of Intuition: Immediacy Sensing-Relationships and Reason. *New Ideas in Psychology* 15: 185–194.

Descartes, Rene. 1971. *Discourse on Method and the Meditations*, trans. F. E. Sutchiffe. London: Penguin Books.

Descartes, Rene. 2004. *Dika'er Sibian Zhexue* [Speculative Philosophy of Descartes], trans. Shang Xinjian and others. Beijing: Jiuzhou Chubanshe. Beijing: Jiuzhou Press.

Einstein, Albert. 1976. *Aiyinsitan Wenji* (Diyijuan) [Collected Works of Einstein (Volume 1)], trans. Xu Liangying and others. Beijing: Shangwu Yinshuguan.

Fan, Xiao. 2003. Guanyu Yuyanyu Siweide Guanxi Jiqi Xiangguan Wenti [On Language, Thinking and Related Questions]. *Yuyan Kexue [linguistic Sciences]* 6: 73–85.

Fu, Jizhong. ed. 1984. *Zhexue Dacidian Luojixuejuan* [A Grand Dictionary of Philosophy, the Volume of Logic]. Shanghai: Shanghai Cishu Chubanshe.

Hong, Qian. 1964. *Xifang Xiandai Zichanjieji Zhexuelunzhu Xuanji* [Selected Works of Modern Philosophy by Western Bourgeois]. Beijing: Shangwu Yinshuguan.

Li, Youliang. 1999. Shijueyixiang: Zhijueyu Siweide Qiaoliang [Visual Imagery: The Bridge Between Perception and Thinking]. *Jinyang Xuekan* [Academic Journal of Jinyang] 4: 68–71.

Piaget, Jean. 1981. *Fasheng Renshilun Yuanli* [Principles of Genetic Epistemology], trans. Wang Xiantian and others. Beijing: Shangwu Yinshuguan.
Polanyi, Michael. 1969. *Knowing and Being: Essays by Michael Polanyi.* London: Routledge.
Qian, Xuesen. 1983. Guanyu Siwei Kexue [On the Science of Thinking]. *Ziran Zazhi* [Chinese Journal of Nature] 8: 563–567+572.
Qian, Xuesen. 1986. *Guanyu Siwei Kexue* [On the Science of Thinking]. Shanghai: Shanghai Renmin Chubanshe.
Saussure, Ferdinand de. 1980. *Putong Yuyanxue Jiaocheng* [Course in General Linguistics], trans. Gao Mingkai. Beijing: Shangwu Yinshuguan.
Wang, Lu. 1999. Lun Birande Dechu [On the Concept of 'a Conclusion Inevitably to be Drawn']. *Zhexue Yanjiu [philosophical Research]* 10: 51–59.

Chapter 10
The Linking Matters: An International Poetics of Sense-Making and Innovation

A subtle chain of countless rings.
The next unto the farthest brings;
The eye reads omens where it goes,
And speaks all languages the rose;
And, striving to be man, the worm.
Mounts through all the spires of form. (Emerson 2009, p. 18)

—Ralph Waldo Emerson.

Charles Bernstein has been, of all contemporary American poets, the one who has done the most to bring back those important words and phrases that tend to be "excluded" from circulation. (Perloff 2021, p. 86)

—Marjorie Perloff.

When a mother gives an egg to her child and says "egg", she is helping her child establish a "link" between language and the world. But what is the nature of this link? As the great Swiss linguist Ferdinand de Saussure explained, *langue* (by which he meant particular languages: like French, or English, or Mandarin Chinese) is a system of signs which parcels out the world of sense into discretely sayable things. The signs that comprise this system are complex: they are composed of a *signifier*—the acoustical image, as he called it, that is formed from a combination of a given language's phonemes—and a *signified*—the conceptual image or item recalled and indivisibly linked with that string of phonemes. The linguistic sign is not the thing in the world which it names (Saussure 1959, pp. 65–66). In the scene described, the mother teaches the child the link between all three elements in a single stroke, bringing the child irrevocably into the world of language—both the particular language, through which this introduction is made, and language in general, what Saussure called *langage* (the fact of human speech). Eventually, the child will learn to draw a self-conscious distinction between language and the world which it denominates, between what is sometimes called the linguistic functions of *use* and *mention*, as when he is able to *refer* to "egg": not the reproductive ovum

Z. Li and D. Braun, *China Being Led and Leading*,
https://doi.org/10.1007/978-981-96-6803-8_10

and its nutriment, but the three-letter word spelled */e/g/g/.* The creation of these links is the foundation for all human thinking, upon and out of which all of our most complicated thoughts are built. It is from the perspective of these links that we can examine some of the most pressing questions concerning what we will call in this chapter, *international poetics*, the communication of innovations and norms within and between the poetry of particular languages and cultures, and beyond.

The best recorded story to demonstrate how the first links between a signified, a signifier, and the real world are created is that of Helen Keller. In a classical instance from her memoir she recalls, "as the cool stream gushed over one hand, she spelled into the other the word water, first slowly, then rapidly. I stood still, my whole attention fixed upon the motions of her fingers. Suddenly I felt a misty consciousness as of something forgotten—a thrill of returning thought; and somehow the mystery of language was revealed to me. I knew then that 'w-a-t-e-r' meant the wonderful cool something that was flowing over my hand. That living word awakened my soul, gave it light, hope, joy, set it free!" (Keller 1903, p. 23) In this passage, Keller vividly describes the moment when "the link" between the word "water", and that material, wonderful cool flowing water of the world that was impressed upon on her mind. Though the signifier of the linguistic sign is objective, common to all speakers of a language, the cognitive image to which it is linked, and the emotional associations it bears, is personal, subjective and changeable. The significance of this division between the objective elements of language, and the subjective half to which they are bound and through which they are supported, is significant to the study of poetics.

T.S. Eliot's theory of an "objective correlative" is a case in point. He states: "The only way of expressing emotion in the form of art is by finding an 'objective correlative'; in other words, a set of objects, a situation, a chain of events which shall be the formula of that particular emotion; such that when the external facts, which must terminate in sensory experience, are given, the emotion is immediately evoked" (Eliot 1920, p. 92). Eliot is correct in suggesting that such an emotion can be "evoked". However, it is a mistake to assume that "the link" between "the evoker" and "the evoked" is objective. In fact, as exemplified in the case of Helen Keller above, the emotional association with the image of a particular object in one's mind is formed by a combination of personal experience and collective instruction, the results of which are at once common enough to allow communication among speakers, yet irreducibly individual, and variable among one another, such that we can never know if our signifieds are identical to each other's.

This is indeed one of the great mysteries and miracles of language. Beyond the brute difference of other minds, there are the idiosyncrasies and vagaries of experience that contribute to the formation of our sense of our language. For Helen Keller, the emotional response evoked by the word "water" included the unique joy and enlightenment she experienced when she learned the word. The word retained for her a sense of the discovery of its link to the world. Another reader, one perhaps not deprived of their senses in the way Keller was, might have a completely different emotional response to "a set of objects, a situation, a chain of events," not only with respect to other readers, but to the artist endeavoring to evoke the objective correlative itself.

Yet people do share certain common experience, which is what makes Eliot's theory possible in the first place. All human beings, being human, share certain life experiences and outlooks upon the world that enable them to enjoy the same literary works. The notion of a classic work, enjoyed by people of all nations around the world, is tacit proof of the commonalities across regional differences that make international literary and artistic success possible. People of the same national or cultural background will of course share more personal experience than those of different national or cultural backgrounds. There are artistic works that are highly favored in one culture while not well regarded in others. A good example is the novel *A Dream of Red Mansions* (《红楼梦》, 1744–1754) which is regarded as the best novel ever written in Chinese, yet hardly read in the west.

Literature often serves a pedagogical function. The degree to which works are read, and continue to enjoy success, often depends on their ability to continue to teach readers something about themselves, and their world. Ezra Pound, another remarkable theorist of literature (and poet), is among the most vociferous exponents of this theory of literary efficacy. He vividly describes the rewards a fruitful reading experience offers as "that sense of sudden liberation; that sense of freedom from time limits and space limits; that sense of sudden growth, which we experience in the presence of the greatest works of art" (Pound 1968, p. 4). Indeed, rewarding reading experiences are those that enlighten and develop our minds, stimulating them to great satisfaction. At the root of any literary judgment is in fact this affective dimension of reading, which serves as the motive force by which literary traditions are sustained. To read in an engaged way, to persist in reading, is to be somehow moved by it.

How does this affective core of reading, which binds traditions and communities together in spite of their differences and distances, hold for the international communication of modern poetry, not least when even the most highly experienced reader of one culture can feel lost in the works of another? Bewilderment occurs not just with respect to the idiomatic sense of another language and culture but, when it comes to poetry especially, the ways in which an artist plays and puns with every level of that language. A story told by the distinguished scholar Huang Yunte about his colleague Zhang Ziqing, is illustrative. Reading the American poet Charles Bernstein's poem "Fear of Flipping," Zhang persistently asked the poet for the lexical meaning of the words in the poem. Huang explained, "the poet is more invested in the ring of echoes of *wall, ball, fall, all*, and even the half- rhyming *repel*, than the lexical meaning of these words. The ricochet of sounds and syllables, creating the titular fear of flipping, like a flip or slip of tongue, looks to walls to keep it inside or floors to hold it up" (Yunte 2021, p. 275). In other words, Bernstein is experimenting with the sonic dimension of poetic lines; indeed, one could say that the 'meaning' of his verse here is produced by his play with the reverberation of rhyming syllables across the poem. Poetic meaning is therefore not restricted to, or even primarily, lexical here. The title of the poem sets the terms for this play by punning on the phrase "fear of falling," a substitution of one term / phobia for another, which *flips* the sense of the original phrase on its head. The echoing internal rhymes create a verbal image which gives shape and body to this gesture of flipping, retaining the ghost of the original phrase even as it ricochets across the altered soundscape of the lines. This practice will no doubt be recognizable to readers who are familiar with the poetics of the Language School, an avant-garde movement Bernstein helped to found. The play serves as a framework for linking mind and world beyond and between the confines of individual languages, and is definitive of Bernstein's practice.

Huang Yunte's interpretation is not difficult to understand. However, it was wholly foreign to Zhang Ziqing, and would almost certainly be to anyone who did not come to Bernstein's work with the framework of sound and cognitive play already in mind. Modern poetry like his is not unique in being theory-laden—that is, constructed and expounded according to the unique poetics of its practitioners. Nevertheless, modern poetry and poetic theories are two sides of a coin; they stand by working together—all the more so as poetry becomes esoteric in form, further removed from the conventions of ordinary language use, and governed increasingly by rules of composition unique to it. Without knowledge of the theories which govern such an esoteric art, one can find oneself at sea while reading a modern poem.[1] This is especially true of poetry where innovation does not occur at the lexical level either: indeed, where the *poetry* at stake is not a matter of lexical play. The divergence of modern poetry from the rules which governed previous traditions—rules of a more subtle kind of artifice intelligible to a broader literate class—has made the dissemination of its doctrines and theories a necessary part of its reception and interpretation. The difficulty a lay but native reader faces with work like Bernstein's is exacerbated in the international context, where neither fluency in the language of composition, nor education within a broad concept of the originating culture, can serve as sure guides. It is paramount that Chinese scholars introduce both modern poetry and modern poetic theories together, teaching them as two facets of the same literary phenomenon.

Many modern poems make good sense in a lot of ways other than the traditional lexical one, which is why they can seem quite difficult to understand. T. S. Eliot once wrote: "We can only say that it appears likely that poets in our civilization, as it exists at present, must be *difficult*. Our civilization comprehends great variety and complexity, and this variety and complexity, playing upon a refined sensibility, must produce various and complex results. The poet must become more and more comprehensive, more allusive, more indirect, in order to force, to dislocate if necessary, language into his meaning" (Eliot 1954, p. 289). Indeed, modern poetry is as difficult to comprehend as modern civilization. The difficulties are comparable, mimetic even, in so far as the poet is driven, in Eliot's reasoning, by a vocational maxim to both reflect and train the sensibility of his audience to his work through the dislocations of language he performs. The difficulty of modern poetry is a difficulty inherent in its context: modern civilization. What of its value? The other aspect (ever present) of Eliot's judgment. A difficult poem is good not because it is difficult. In fact, the difficulty of many poems is not that difficulty of modernity refracted, but rather a failure to adequately make sense of the incoherence the poet intuits in his era. It is a subtle difference, one with which Eliot was principally concerned. A difficult

[1] See (Li 2020, pp. 27–34) In this essay, Li argued that modern poetry and modern poetic theories have to be read side by side to make good sense of both of them.

poem is good only when it creates one more possibility of "forcing language into meaning," in an unconventional way. Again, "Fear of Flipping" is exemplary. From the perspective of linking, it is an exploration of more possible ways to make new, and possibly more efficient, thought ways and patterns within *and* across cultures. The poem's difficulty is likewise a function of the way in which it is approached. Though Eliot would demur to such a consequentialist proposition, perhaps the test of a difficult poem's quality may be the very satisfaction of mind, its inspiration and development, that has affectively and cognitively bound generations of poetry readers to one another in a tradition millennia-old, and world-wide.[2]

What then of the transposition of these difficult poems into foreign contexts. From one perspective, it would be easy to conclude that poems like "Fear of Flipping" simply cannot be translated into Chinese. Semantics are not what a translator ought to target here, yet there are no characters in the Chinese language that reproduce the poem's soundscape either: *wall, ball, fall, all*, and *repel*, are constructions of the sound system of English. Chinese phonology simply does not permit their formation. Yet this perspective is impoverished, for the link the poem creates (the link which is its essential, creative practice and energy) is certainly "translatable." The poem's signature effect, its 'fear of flipping' so to speak, can be reached in the target language of Chinese, and the minds of readers from this or another culture, like those of its author and his native culture, can be enlightened and developed by a translation which aims to *translate* those effects. From the linking perspective, the reward for reading a poem is to build up some new and better links, so that the minds of its readers can grow.

In bringing, i.e., "translating", poems like "Fear of Flipping" to readers in China, we need to explicate them in detail, line by line, giving more detailed interpretations than what Huang Yunte does in his essay; but we also must consider the general theory and framework of mind that the poem conjures. For it is only by doing both that Chinese readers will be rewarded in their encounter with the difficulty of works like those of Bernstein, or his Language School peers. This is the true project and mission of translation.[3] To deal with such poems that stand closely with the linguistic features of the particular language in which they are written that cannot be replicated in Chinese, the strategy for translation is not to focus on the technical details of linguistic features, but on helping readers in China in understand the ways of the poem, i.e., the frameworks of mind presented, so that they can not only understand themselves, but create links in Chinese in the same spirit, replicating these features where possible, according to the rules of their native tongue.

Marjorie Perloff has noted: "Charles Bernstein has been, of all contemporary American poets, the one who has done the most to bring back those important words and phrases that tend to be 'excluded' from circulation" (Perloff 2021, p. 86). In

---

[2] See (Li 2021, pp. 1–8) In this essay, Li argues that one major function of modern poetry is to promote the development of human's intellectual and emotional capacities.

[3] In fact, this is what we have done in our on-line bilingual course on modern poetry in English. This on-line course can be reached at: https://www.ulearning.cn/course/25598. In this course, Charles Bernstein is invited to have given a talk on an excerpt from *Dark City*, in which he gives a line to line interpretation. This is indeed the best way to "translate" a difficult modern poem.

other words, the contribution Bernstein's works have made is not only to serve an individual reader by promoting his/her intellectual and emotional growth, but also, and more importantly, to serve contemporary American language and culture as a whole. With poems like "Fear of Flipping", Bernstein constructs and reconstructs delicate links that promote the growth of the contemporary American capacity for thought. That is to say, his work has contributed to the growth of the thinking capacity of the American cultural being which, if well "translated," can help other cultural beings develop in similar, relevant areas too.

Different from lyrics, narrative works, both in verse and prose, tell stories that define the formation of certain links, as well as the associated emotions, so that they can often be translated in a traditional way. Story travels across cultural borders much more freely than poetic technique.

In the field of international cultural communications, a mind, or a culture at large, grows in two ways: one is of transplantation, the other of inspiration. The key difference between these two learning ways is that the former offers something that cannot be logically developed out of the exercise of the learner's own mind, or the recipient culture's institutional self-renewal, while the latter brings something that can be logically achieved by the recipient person or culture.

A few further examples from various contexts should demonstrate the difference. When Sun Yat-sen established the Republic of China in 1912, he attempted to "transplant" the whole American political system into China. The effort was a failure because the system could not function well within Chinese culture. After the May Fourth Movement in 1919, the western ideology of free love and free marriage were introduced into China, which inspired many young people, who totally understood them, cherished them, and were willingly guided by them. A corollary intuition of these values already existed in Chinese history: been many people fought for their freedom of love and marriage, without necessarily having developed a theory of these practices. In some cases, the transplantation model and the inspiration model have been combined together. One such example is Socialism with Chinese Characteristics; Socialism was transplanted from the former Soviet Union, while Chinese Characteristics refers largely Chinese people's own innovation, partly on the basis of traditional Chinese political practice, and partly inspired by Western political practices.

When a Chinese student learns English, s/he needs to learn a vocabulary and a grammar / syntax (words, and the rules for their formation and combination). In this way, his/her mind grows by "transplantation". When a Chinese scholar learns Charles Bernstein's poetics, acquiring a totally new way of thinking, it is also of transplantation. Inspiration, by contrast, is the event of learning something that can be interpreted, understood, and made good sense of in the context of one's already established knowledge. For example, all traditional western poetry, especially the works of the Romantics, such as those by William Wordsworth, can be easily understood by Chinese readers, as they share much of the spirit with traditional Chinese poetry.

Where do these processes of transplantation and inspiration fit in the current world of international poetics? Among the most interesting instances in the communication of inspirational learning is what one may call mis-interpretive innovation. These are cases defined by a fortunate mistake, in which the application of a norm in the target language and culture to the translation of a work produces something incongruous with the original cultural perspective. One famous example is Ezra Pound's invention of the "Ideogrammic method". As Xie Ming has written, "this etymological, compositional theory of the ideogram, from which Pound derived his 'ideogrammic' method, had an enormous impact on his thinking about poetry and other cultural matters, and on the writing of the *Cantos*" (Xie 1999, pp. 236–237). The method has influenced many poets in the west: "an American mind, brought to ideographs by an art historian of Spanish descent who had been exposed to Transcendentalism, derived Vorticism, the Cantos, and an 'ideogrammic method' that modifies our sense of what Chinese can be" (Kenner 1972, p. 162). Indeed, it is for this reason that Pound is said, in a well-known idiom, to have 'invented Chinese poetry in English". And yet, as explained on the back cover of the book *The Chinese Written Character as a Medium for Poetry*, the Chinese language is just a set of signifiers, like the English language, or any other language (Fenellosa 1936, Back Cover). This now seems to be common knowledge to most English readers. However, Ezra Pound's invention of the "Ideogrammic method" made perfect sense in its context, and it was a wonderfully productive method for the composition of his works, as well as those of many other poets working in English. It was an extremely valuable invention in poetics in English per se, inspired indeed.

There are many more examples of mis-interpretive innovation. Let us offer a personal one. When Zhimin Li was invited to give a talk on Ezra Pound's lyric "The River-Merchant's Wife: A Letter", Li found the other three panelists interpreted the poem as a war poem, the context for which was England's involvement in World War I.[4] This appeared quite strange to him, as Li had been accustomed to interpretating the poem in the cultural context of its original author, the great Chinese poet Li Bai (701–762). In the traditional Chinese cultural context, this poem is normally taken as a love poem that romanticizes the mutual love and devotion of a young couple, which is considered a key virtue of the family ethics highly valued by Confucianism. Li found his American colleagues' new interpretation compelling, making good sense as it does of the historic context in which the poem was translated and read in England. In fact, this new interpretation is inspiring and fascinating, and has contributed to the growth of Li's understanding of the original and translated poem. What we can see from this example is that poetry not only exists in its original or translated context, but between them, in the historical and cultural rhymes that mutually illuminate diverse regions of the literary tradition.

Let us give another example that illuminates the importance and shifting influence of context. A famous Chinese scholar, Yue Daiyun, once held a seminar to discuss the novel "Xiao Erhei's Marriage (小二黑结婚)" with her American students. In the novel, there is a character named Sanxiangu (三仙姑) who often makes herself up to

[4] See an on-line conversation "A discussion of Ezra Pound's "The River-Merchant's Wife" among Al Filreis, Emily Harnett, Josephine Park and Zhimin Li.
https://media.sas.upenn.edu/app/public/watch.php?file_id=208367

look more beautiful, which is meant to be inappropriate, as she is of the working class. The conventional comments of the proletarian literary circle in China concerning this character are always negative. However, Yue Daiyun found all her American students supported Sanxiangu, as they thought there was nothing wrong with her making herself up. On the contrary, they considered Sanxiangu to be an admirable woman, as she seemed to them to love life (Yue et. al. 1989, p. 19). Yue Daiyun came to agree with her American students' comments, and has been disseminating their views ever since to her students and colleagues back in China, which is surely a contribution to the interpretation of the character Sanxiangu as well as the whole novel in China.

The purpose of international interactions is not to make all cultures the same. Rather, international interactions can make all parties more perfect in their own way. We learn from each other in the transplantation model only when there is no alternative. We apply the model of inspirational learning in most cases. The overriding principle to decide whether any international communication is fruitful or not is whether it makes good sense in terms of the recipient individual or culture, indeed, whether it enriches the recipient through the change it rings.

In the model of inspirational learning, the exchange can move in both directions: the innovative knowledge produced by the recipient may depart from the codes and conscience of the original culture, and yet in doing so inspire something novel in return, within the original culture. With the back and forth of such international communications, human knowledge on the whole is greatly expanded. In fact, the method of international communication, especially of the mind-expanding forms of poetry, is perhaps the best way for humanity to develop itself by diversifying itself: that is, to resist the pull of sameness.

International interaction follows more or less the same principles in other fields. For example, in the field of politics, China and the West have learned and benefited from each other, and will continue to do so in the future. Jacques Gernet has written: "China furnished the first example of a disciplined, rich, and powerful state which owed nothing to Christianity and seemed to be based on reason and natural law. It thus made a powerful contribution to the formation of modern political thought, and even some of its basic institutions were imitated by Europe" (Gernet 1982, p. 523). Indeed, Gernet convincingly argues that what the West has learned from China it has learned in the inspirational model. In return, China has learned a lot from the West as it developed during the modern age, much of which has transformed Chinese society to a great extent, such as in the fields of education, industrialization, urbanization and so on. And again, perhaps in the future, some of modern China's successful institutions might serve as good examples from which the West might learn, and so on in perpetuity (Li 2012, pp. 111–127).

Ralph Waldo Emerson believed that all forms of life are linked: "A subtle chain of countless rings/ The next unto the farthest brings" (Emerson 2009, p. 18) as he writes in the poem which prefaces his essay "Nature." Helen Keller's story about the creation of the link between the signifier "water," the concept water signified on her mind, and the water out there in the world, is a story about the origination of thought, without which she would have lived in a kind of intellectual darkness all her life. But the story is also general: if human beings could not create links between the world and the world of signs, human beings would have lived in the darkness as well. Without poetry to further enhance these links, or to break and remake them, and without its transposition between languages, in which those links are once more remade into a monster of linguistic and cultural confusion (in the etymological sense of this word), our thought would be even darker. The linking is everything; it is, as Emerson reminds us, life itself.

Works Cited

Eliot, T. S. 1954. The Metaphysical Poets. In *Selected Essays*, ed. T. S. Eliot, 281–291. London: Faber and Faber.

Eliot, T. S. 1920. Hamlet and His Problems. In *The Sacred Wood: Essays on Poetry and Criticism*, ed. T. S. Eliot, 87–94. London: Methuen.

Emerson, R.W. 2009. *The Essential Writings of Ralph Waldo Emerson*, ed. Brooks Atkinson. New York: The Modern Library.

Fenollosa, Ernest. 1936. *The Chinese Written Character as a Medium for Poetry*, ed. Ezra Pound. San Francisco: City Lights Books.

Gernet, Jacques. 1982. *A History of Chinese Civilization*, trans. J. R. Foster. Cambridge: Cambridge University Press.

Huang, Yunte. 2021. Ten Plus Ways of Reading Charles Bernstein: Improvisations on Aphoristic Cores. *Boundary 2: An International Journal of Literature and Culture* 4: 255–278.

Keller, Helen. 1903. *The Story of My Life*. New York: Doubleday.

Kenner, Hugh. 1972. *The Pound Era*. London: Faber and Faber.

Li, Zhimin. 2011. *Zhishi de Shanyuzhen* [The Good and the True of Knowledge]. Beijing: Renmin Chubanshe.

Li, Zhimin. 2012. The One Way Model of Cultural Interaction: Literary Interactions between China and Cambridge. *The Cambridge Quarterly* 41 (1): 111–127.

Li, Zhimin. 2020. Lilun Zhudaoxia Yingyu Shige de Xiandai Zhuanxing [Modern English Poetry: Innovation through Theory]. *Waiguo Yuwen Yanjiu* [Foreign Language and Literature Research] 5: 27–34.

Li, Zhimin. 2021. Xiandai Shige de Chuangxin Jingshen [Innovative Spirit of Modern Poetry]. *Waiguo Yuyan Yu Wenhua* [Foreign Languages and Cultures] 2: 1–8.

Perloff, Marjorie. 2021. Introduction to Charles Bernstein's Distinguished Wenqin Yao Lectures at Zhejiang University, Hangzhou, Fall 2019. *Boundary 2: An International Journal of Literature and Culture* 4: 85–90.

Pound, Ezra. 1968. A Retrospect. In *Literary Essays of Ezra Pound*, ed. T. S. Eliot, 3–14. Toronto: George J. Mcleod Ltd.

Saussure, Ferdinand De. 1959. *Course in General Linguistics*, trans. Wade Baskin, ed. Charles Bally and Albert Sechehaye. New York: Philosophical Library.

Xie, Ming. 1999. *Ezra Pound and the Appropriation of Chinese Poetry: Cathay, Translation, and Imagism*. New York and London: Garland Publishing.

Yue, Daiyun, et al. 1989. Nvquan Zhuyi yu Wenxue Piping [Feminism and Literary Criticism]. *Wenxue Ziyoutan* [Free Talks on Literature] 6: 18–24.

Chapter 11
Truth as a Construct and the Growth of Knowledge

In traditional China, people were born into Confucianism and the imperial system, taking each for granted. They did not question whatever was given to them, assuming that the way of life proposed must be followed, without questioning the system of rewards and punishments articulated by it.

However, starting from the Opium War launched by Great Britain in 1840, China was repeatedly defeated by the Western powers, which opened people's minds in China to an alternative to both Confucianism and the imperial system. In the Sino-Japanese War of 1894–1895, China was disastrously defeated by Japan, a country considered by China to be an inferior nation in all aspects, and which had yet learned from the West to become more powerful, radically undermining the Chinese confidence in its traditional systems. As a result, people in China overthrew the Imperial System in the Xin Hai Revolution in 1911, openly introducing Western ideologies into China as a replacement for Confucianism during the May Fourth Movement in 1919, which took *Science* and *Democracy*, two key concepts of the Western culture, as its slogan. Ever since, China has departed decisively from both Confucianism and the imperial system, and embarked on a road toward a new nation, of which the maturated stage is envisioned to be the modernization of all fields, a consensus that has been held for many years now.

In traditional China, people simply followed Confucianism and the imperial system without bothering themselves by asking such questions as what the advantages and disadvantages of Confucianism and the imperial system were, whether there were better alternatives, and so on. However, after they were presented with a way of life found in the West, a way that seemed to be much better, they began to think carefully about such questions, and make alternative choices. They began to make those choices during the Xin Hai Revolution in 1911 and the May Fourth Movement in 1919, which took the most simple form of transporting everything from the West. However, it was extremely difficult for many people in China who had been used to Confucianism and the imperial system to understand the Western cultural values in their true senses, and it is even more difficult for people in China

Z. Li and D. Braun, *China Being Led and Leading*,
https://doi.org/10.1007/978-981-96-6803-8_11

to follow them in their life practice. Indeed, radical cultural transformation is among the most complicated challenges for a nation, especially a large nation with a long history of traditional ways.

Since it did not work well by simply transporting everything from the West, people in China had to work out their way to modernization, searching for truth in both the material and human worlds in the hope to find the best path for China to survive and thrive in the modern age.

In this chapter, it is proposed that there are actually different kinds of truths, including exclusive truth, such as those in *Science*, and construct-truth, such as those in *Democracy*. This chapter puts forward a theory of construct-truth to demonstrate how to search for it in the realm where it is applicable—the arts and humanities, and the social sciences—as well as how to promote the growth of knowledge relevant to individuals and society in general.[1] This can provide a solid theoretical framework to help people during their way to modernization.

Defining Construct-Truth

Truth is often taken as something exclusive, such as in the equation "$1 + 1 = 2$" that is a piece of *exclusive truth* because there is only one correct answer "2" to "1 + 1." Any other answer is incorrect. In natural science, there are abundant examples of exclusive truths that normally contain just one proposition and one correct judgement.

In addition to the exclusive truth, there is another kind of truth that can be named *construct-truth*, a form of truth that is not exclusive. Such a truth is a kind of construct that is made of a set of ideas. Its truth does not lie in whether any idea included is right or wrong, better or worse, but in the actuality of the dynamic construct formed by all the ideas in a certain field. Unlike exclusive truths, construct-truth widely exists in the fields of arts and humanities as well as the social sciences, where human beings are the object of studies. Construct-truth is significant in helping us not only to understand

---

[1] Construct-truth is English translation for a Chinese term "格局式真理" (*gejushi zhenli*) coined by Zhimin Li. The original Chinese term "格局" (*geju*) refers to (1) a situation in which different forces, members or ideas are competing or not competing with each other, (2) a cultivated worldview that is broad-minded and noble rather than narrow-minded and low-graded. Zhimin Li hesitated a little while before he decided to pick up the English term "construct" as the translation for "格局" (*geju*) as the word "construct" might cause readers to relate the theory of the construct-truth with structuralists theories invented in the West. Indeed, Li agrees that he is academically indebted to his reading experience of various structuralists works. However, the term construct-truth is invented for its own purposes, which is independent of any structuralist theory. A reader is not supposed to read any structuralist theory in order to understand the theory of construct-truth proposed in this chapter. In fact, one of the main purposes for the invention of this term construct-truth is to counterbalance the stereotyped mindset nurtured by the philosophy of the excluded truth that has left much impact on structuralism. Besides, the theory of construct-truth emphasizes on the subjective efforts to putting ideas together (as suggested in "con") to make more sense of the world, i.e., understanding the world in a complete picture and initiating more sophisticated thoughts rather than discovering already existed objective structures in the world.

the overall situation of a field or a society, but also to be independent of the mindset to search for and rely on the exclusive truth. For instance, some readers say Li Bai's (李白) poetry is better than Du Fu's (杜甫), while others say the opposite; some say both of their poetry is good, while others say neither is. In this case, the exclusive truth does not apply, since none of the four statements is exclusively correct. Rather, construct-truth applies, as the four ideas made by the four different groups of readers form a construct that is true of the group in itself.

By searching for the exclusive truth, we can promote the development of natural science, while we can promote development of arts and humanities as well as social sciences by working on construct-truth. The former is object-centered, while the latter is human-centered, taking humans as its object. If we call the exclusive truth hard knowledge, then we shall call the construct-truth soft knowledge. The former is rigid and must be obeyed strictly, while the latter is flexible and offers many different choices.

There are different sorts of construct-truth of different scopes. A table is drawn below, for the convenience of more detailed explication, of the most typical type of construct-truth, i.e., that of a society (Fig. 11.1).

In the table above, the black dot (A) represents the exclusive truth, the white dot (C) represents the situation when no new ideas are produced and no knowledge is grown, and the segment (B) between A and C represents the construct-truth. B is divided into three kinds of construct-truth: b1 represents the basic consensus of a society, b2 represents the procedural consensus of a society, and b3 represents the budding period when all different ideas are being created and developed, competing or not competing with each other, in coexistence in a society. Both b1 and b2 evolve from b3, which serves as origin and base of the former two. The views in b1, b2 or b3 form a kind of construct-truth in their respective fields, and all the views in b1, b2 and b3 together form a most inclusive construct-truth of a society.

The basic consensus of a society is very close to the exclusive truth, as it is deeply believed in by the majority or even all people of the society. It is a kind of quasi-hard knowledge. However, it is not ultimately hard knowledge, because it remains context specific, and is thus not universally true. For example, there has been a very deeply rooted belief that says "one who kills another person pays with one's life," which has been a piece of basic consensus in China for thousands of years. However, it is neither absolute nor universal. For example, one is not considered to be a criminal,

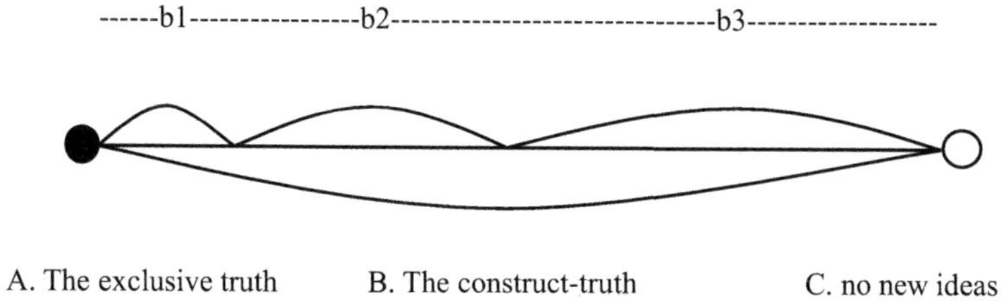

Fig. 11.1 (A) The exclusive truth (B) The construct-truth (C) no new ideas

nor to be punished, if one kills another person in self-defense. Besides, more and more nations in the modern age have abolished the death penalty, and more and more people in China have started to doubt it too. In due course, the belief of "one who kills another person pays with one's life" might be disqualified as a piece of the basic consensus in China.

Unlike the exclusive truth, there are always different ideas that co-exist with a piece of basic consensus. For example, there were, are and will always be people, Buddhists for instance, who disagree with the belief that "one who kills another person pays with one's life" in China. In fact, each belief in the basic consensus, being prominent, coexists with different ideas, forming a mini type of construct-truth. It should be noted that the prominent ideas related exchange places with the covert ones in different societies or ages. The exclusive truth, by contrast, always remains the same at all times in all places.

Procedural consensus is an agreement, a principle, a law or the like that is established via a certain procedure that is unbiased and generally agreed upon by the public. Procedural consensus is lower in the degree of consensus than basic consensus. It can be something supported by the majority of a society, or not in certain cases. For instance, when Barack Obama's health care revision bill was passed and became law in 2010, it was supported only by a narrow majority of 219 votes to 212 in the House. When Donald Trump became president of the USA as the result of the procedural consensus in 2016, he lost to Hillary Clinton in terms of overall votes.

The key to reach a piece of procedural consensus is to establish certain procedural rules in advance that are neutral to all participants. To do so, there should be an independent third party established by certain neutral legal rules, who is authorized to ensure the procedural rules are strictly observed, and the result to be recognized by the public. It should be noted that the consensus reached via procedural rules, i.e., the procedural consensus belongs to b2, while the neutral procedural and legal rules belong to b1, i.e., the basic consensus of a society. If a society does not have certain piece of basic consensus (b1), the related procedural consensus (b2) could never be reached.

Procedural consensus is very important for the good functioning of a society, as it enables the whole society to get people who differ in ideas united in action to move forwards. In some cases, something must be done to respond to a situation, while it is impossible for the whole population, or even the majority to reach consensus. In such cases, the procedural consensus can be applied to allow the society to take action. Sometimes procedural consensus might not produce the best way to respond. However, the very act of united response is good enough in itself. For instance, when a house is on fire, the most urgent thing is to put it out, not to wait to argue by whom or how it should be put out. A society without a mature system capable of producing procedural consensus, with certain neutral procedural and legal rules enshrined, might fall into chaos in certain critical situations, through endless internal disputes or even internal fights. In fact, China has fallen into such disastrous situations before and may well again, suffering great losses in both wealth and human lives. Indeed, China is in urgent need of establishing such systems to produce procedural consensus to deal with issues that have turned up, or will turn up, so that it won't fall into the same disasters of disagreement as it has in the past.

One finds most ideas in the budding period as shown in the table. With the passage of time, some ideas are proven to be unworthy, and are given up as a result, while others are proven to be worthy, improved and accepted by more and more people, and might even become part of the basic consensus.

The budding period is the place where ideas germinate and develop, and it is crucial to the development of knowledge in a society. In order to facilitate the production of good ideas, it is necessary to provide fertile soil during the budding period: this might come in several forms, including by providing qualified education to the public, granting freedom in communicating ideas, providing forums for different ideas to compete and complement with each other, and so on. Debates at all levels in all fields should especially be encouraged, as it is not only the best way to promote the birth of original ideas, but also the best way to get the public interested in, educated by, and in agreement with respect to certain ideas. How dynamic the budding period of a society is, including the fact of how much qualified knowledge it produces and promotes each year, is a critical criterion to judge the quality of a society in the modern age.

If people stop producing new ideas in a society for some reason, the whole society may cease to grow in its knowledge base. As a result, the whole society may become rigid, fragile, decayed and even teeter toward collapse.

Why is There Construct-Truth?

Natural science searches for the exclusive truth that governs the material world and makes use of it to serve humanity. The exclusive truth in natural science is often expressed in absolute rules such as those in mathematics, geometry, physics, chemistry, astronomy and so on. Many absolute rules are expressed in formulas such as Isaac Newton's second law "$F = ma$," Albert Einstein's theory of special relativity "$E = mc2$." It is unknown to humanity why such miraculous rules exist, without which, the whole universe could fall into chaos immediately.

The most fundamental reason for construct-truth that exists widely in arts and humanities, and social science, lies in the flexibility of human beings as well as different viewpoints presented by different individuals.

Human beings are able to adapt to different external conditions and produce different ideas that are the premise for the formation of the construct-truth. For instance, one can survive in either hot summer or cold winter, which offers one opportunities of different experiences to produce different ideas. If human beings can only survive in exactly 20 degrees Centigrade, they would not have such opportunities. The more flexibility one has, the more opportunities one is offered, and so the more sophisticated a construct-truth can be. A single idea reflects just one aspect of the picture, thus being impartial. However, the construct formed by all the ideas related reveals the whole picture, which is qualified in the truth construct.

The flexibilities of both a person and the whole of humanity are limited, so is the freedom of their choices and the ideas produced, hence construct-truths are limited as well. For example, human beings normally feel hot when the weather reaches 30 degrees Centigrade, and will feel very hot when it reaches 40, and may die if the temperature rises a little more.

The more typical construct-truth is produced by more than one individual, from a small group like a family, to a social organization of various scales, to a nation, until it accedes to the whole human community. The construct-truth in a family is the most simple, while that in the international community is the most complicated. The criterion for us to judge the quality of a construct is how well it allows the production of qualified original ideas.

Even members of a family who love each other may have very different ideas regarding certain issues, which form all sorts of construct-truths. Like the individual, the flexibility of a family as an entity, stretches to a certain degree, and has a limit. Unlike the individual, a family has different members who are physiologically independent of each other, and have different experiences, different interests and different standpoints. To keep a family intact, there should be certain principles that are observed by all its members. For example, all members should love, care and take responsibility for each other. Otherwise, a family might collapse. At the same time, members of a family should be allowed freedom, as much as possible, in pursuing their own interests, and getting their own potentials fully developed, as fully as possible, which will enable members of a family to become the most productive in qualified original ideas. It is a terrible situation when a family is ruled tightly by a tyrannous parent who deprives members of their freedom. As a result, no new ideas are encouraged to be brought out in the family, and it is dead in terms of the growth of knowledge.

All social organizations of various scales, such as a school, a company, a party or the like all share the most basic feature with family, the smallest-scale social organization, which is to balance between observing certain rules to keep the group well organized while allowing its members the most freedom possible, so that qualified original ideas could be produced to promote the organization's growth and improvement. Any social organization can be bettered, from small technical inventions to strategic reformations, if new ideas are encouraged, while one that does not welcome new ideas is doomed to become outdated soon.

All social organizations have their own features, and have their own methodologies for balancing between keeping order and sponsoring freedom. Perhaps the model for all may be found in organizations that are designed for the production of new ideas, such as universities and other academic institutions. The advantage of such organizations is that they can produce abundant qualified ideas in a short period of time. The disadvantage is that some of the ideas they produce are fake ones that they themselves do not really believe or truly understand.

The nation is the most typical organization of human beings at present, which constitutes the most typical society to which the construct-truth explicated above is relevant. As each nation enjoys sovereignty and claims rights in making its own laws and policies, it wields influence over the construct-truths of the whole society, and all the other types of construct-truth within those. Due to the different ways a nation is run, very different outcomes in terms of the production, promotion and application of qualified original ideas often result. Some nations have contributed greatly to the modern way of life, while others have contributed very little.

The international community provides the field for the forming the construct-truth that is the most inclusive and complicated, as it contains all human ideas and knowledge. Throughout history, the constructs of this community have often been torn down and rebuilt, sometimes through violent wars during which many lives were lost. In some cases, a serious plague or other natural disaster has done the same to the construct of the international community. In the past, a nation with its unique culture, or even a whole civilization, has been demolished on earth, a permanent loss for the international community. It is a forever loss of some specific elements of the construct-truth of humanity too.

The construct-truth can form synchronically. Likewise, it can also form diachronically. For instance, one's understanding of Li Bai's or Du Fu's poetry might change a great deal as one grows from a child to an adult, so that the construct-truth of one's understanding of Li Bai's or Du Fu's poetry is diachronic, i.e., it includes all the understanding during all his ages. For another instance, the construct-truth of international relationships, such as that between China and Japan, or China and Russia, or China and America, and so on, includes those both of the synchronic and the diachronic. Without a good understanding of the diachronic, one cannot hope to understand the synchronic, since there are dialectically dependent on one another.

How to Promote the Development of Construct-Truth and the Growth of Knowledge

In traditional China, both the culturally and politically dominant powers, namely Confucianism and the imperial system, did not encourage people to think independently. Neither did they tolerate any challenge, which was often cracked down upon mercilessly. In such an environment, the production of innovative ideas, as well as the development of construct-truths in all fields, especially those in arts and humanities as well as social sciences, were heavily hindered. As a result, there was not much growth in knowledge, indeed, much less than there should have been, during the thousands of years when Confucianism and the imperial system were dominant.

Confucianism and the imperial system were not invented by God and given to China. In fact, both of them came to be during the budding period (b3) in ancient China, and gradually grew into the basic consensus (b1) of ancient China, thus, dominant. Ironically, once they became dominant, they started to wield their powers to repress the emergence of any idea that might challenge their dominance, which is against the right way that allows ideas to fairly compete with each other, so that the better ones win and take place of the worse. This is why China had developed so slowly in the traditional age and lagged far behind the West when they encountered each other in the modern age.

The same story had happened in the West, especially during the medieval years when Christianity took itself as the absolute truth and anybody who brought about an idea that challenged its dominance was severely punished, sometimes burnt alive. This lasted for almost one thousand years, during which the West suffered even more than ancient China did, as neither Confucianism nor the imperial system was as dominant as Christianity. Fortunately, Westerners reformed their religion and liberated themselves from its heavy constraints. Further, they designed a legal system that granted all people freedom of speech, freedom of press and so on, which was intended to prevent any future power from being so dominant as to be able to repress the emergence of innovative ideas. Indeed, this is among the most valuable achievements for the modern age, as it attempted to guarantee the thriving growth of knowledge in a society.

Both China and the West suffered from a monopoly of power, and witnessed suffocation of original ideas and stagnation of social development for a considerable length of history. It has thus reached basic consensus (b1) in the societies concerned that such tragedies in history should be prevented from happening again, though the fruits that different nations have achieved in this regard differ significantly from each other, with some nations having done a great job while others are lagging behind.

Certain procedural consensus (b2) has played a role in the realization of that basic consensus to stop monopolies of power, which is the peaceful transfer of power between different individuals and groups. The most fundamental reason for people to take and keep power is for interest. In the traditional age, there was no such procedural consensus to ensure peaceful transfer of power, so that people had to fight and kill each other for power, which often cost all parties concerned dearly. Once in power, one became afraid of losing power as it often came at the expense of death. As a result, rulers would rather take great trouble to establish their absolute power and implement extreme ways in repressing anyone or any idea that might challenge it. However, both the cost one must pay to take power, and the trouble to keep it, lost their justifications with the rise of the procedural consensus and the peaceful transfer of power. As a result, all parties involved are leading a much more decent and happier life than in the traditional age. The procedural consensus that ensures the peaceful transfer of power has terminated the traditional game of power-grabbing that is characterized by the winner taking all and loser losing all that characterized traditional politics. Indeed, the development has spawned a modern age in which people share power together and compete to become better, in which no oppression, in theory, of the production of innovative ideas is tolerated.

Indeed, suppression of the production of original ideas has been legally removed in many modern societies. And more measures should be applied to promote the production of qualified original ideas, especially those in the budding period (b3). In this regard, there are at least three items that need to be emphasized.

The first measure is to help people, especially those who are working in academic organizations, to be well informed. Without being well informed, one does not have a solid background to produce qualified ideas, and one cannot tell whether the idea one is proposing is original or not. In fact, there have been a lot of ideas produced in this world in all fields, and a lot of ideas that are being produced are not original, as they do not really differ from those in the past. The more informed one is, the more possible it is for one to produce qualified original ideas, and vice versa. This is why the more open a society is, the more qualified original ideas are produced, and the less open one is, the less qualified original ideas it can hope to inspire.

The second measure is to educate people to be able to think logically, independently and critically, all of which together enable original thinking. Anyone can think. However, only people who are well educated are able to think logically, independently and critically. In our daily life, we often think quite casually. However, when serious work is at stake, one must think carefully.

The best way to see whether an idea produced is good or not is to test it with open criticism. A lot of ideas seem to be fine before they are tested with open criticism, and are proven to be unworthy when they are criticized by sharp-eyed critics who serve in the role of doctor in the world of ideas. This is why people who are not confident of their ideas are afraid of open criticism, and those who are confident welcome it. It might be said that the degree to which one welcomes open criticism indicates the degree to which they are confident or not, and whether one's ideas are worth investigating further. For a society that values qualified original ideas, it is essential to protect their critics with law. Without good critics, a society could be occupied by liars rather than innovative and reliable scholars.

At present, the educational system in China is still trapped in the traditional mindset to a certain degree, which focuses on teaching established knowledge without encouraging logical, independent or critical thinking. This has resulted in the fact that China has been lagging behind in the world in terms of innovation in all fields, though its economy has been developing with the greatest speed over the past forty years or so. Once the educational system in China is reformed to be innovation-orientated, abundant qualified original ideas will be produced, and China will become one of the most important centers for innovation in the modern age.

The third measure is the application of the rule of law to regulate all sorts of activities of ideation. In the modern age, freedom of speech and press is granted to every citizen in principle. However, such freedom, as a right for modern citizens, is restricted by other rights. For example, no one is granted the freedom of speech to libel others. In certain cases, it is very delicate and difficult to tell what a libel is, or how serious it is, or whether the law should interfere. However, in most cases it is clear where freedom of speech and press is granted while nobody's legal rights are violated.

In China, there were and are cases in which certain people spread vicious rumors or made personal attacks in the name of free speech. In this way, they made people afraid of or even despise freedom of speech. Thus it is clear that it is important for people's basic rights to be legally protected so that they are not hurt by people's words that are said in the name of freedom of speech. Otherwise, freedom of speech risks losing the support of the general public.

Meanwhile, some basic rules need to be set up to ensure people have fruitful discussion, rather than frequently fall into quarrels and emotional confrontations. In fact, there have been some good academic rules and procedural consensus that could guide how to establish fruitful discussions, which may apply to educated people across any topics.

Many ideas, such as those of traditional Confucianism, the imperial system, extreme religious and political thoughts (those of Adolf Hitler for instance) that have been proven to be wrong or unworthy are still kept in the present day. They form all sorts of construct-truths with ideas that have criticized them and replaced them. They must be treated professionally, always to be mentioned in a construct in which they are properly criticized. The value of their being kept is to provide some lessons for people to learn, having people warned not to repeat mistakes that happened in history, guiding future development.

Conclusion

We have thus discussed the theory of construct-truth from perspectives of what, why and how, which presents a general picture. It is good for its purpose so far, though more work could be done to have it more fully illustrated in the future.

In traditional China, there was no lack of great people, possessed of the virtues of intelligence, bravery, diligence and more. However, it did not occur to them that they could and should liberate themselves ideologically from the monopoly of traditional Confucianism and the imperial system. Fortunately, people in China eventually decided to move on after seeing an alternative exemplified by the West, and have entered the modern age after enormous suffering, courageous fighting, and the loss of many lives.

Undoubtedly, the United States of America has so far been proven to be the most successful state in the modern age, which has contributed the most qualified original ideas. And the most well-known contribution of the United States of America to the world is its unique design of the political construct formed by its system of government, divided between legislation, administration and judiciary, which is in accordance of the theory of the construct-truth put forwards in this Chapter. The value of this truth does not lie in any of the three branches per se, but the construct itself. And now it is time to get people educated to the philosophy that has made the unique design so successful, as is explicated in the theory of the construct-truth that will bring many more successes.

Although China has given up traditional Confucianism and the imperial system for more than one hundred years, and has embarked on the road to a modern state, some people often fall into the mindset that was nurtured for thousands of years by traditional Confucianism and the imperial system. The mindset, as in the way it was nurtured, presumes there is always one truth or authority lying behind everything, which has designed the world accordingly. The same mindset is also quite popular in the West as the result of the long-term monopoly of Christianity during the middle ages. The mindset is not modern, but traditional.

The situation gets worse when the mindset is reinforced by the philosophy of exclusive truth that is popular in the natural sciences, which have achieved great success in the modern age. Many people have, consciously or unconsciously, brought this mindset into understanding, interpreting and initiating issues in arts and humanities as well as the social sciences, which have caused a lot of problems in the modern world. We can call the mindset the "exclusive-truth mindset," which should be carefully avoided when we are dealing with human beings.

People of the exclusive-truth mindset are inclined to think that there is one truth, the one truth, the absolute one, and anything different from it is an error. Such people could become racists, extremists in religious and cultural issues, without even necessarily realizing it. In extreme cases, such people could easily grow to be a Nazi in all issues, and might go to extremes to suppress and even to kill people who are different.

On the contrary, the theory of the construct-truth emphasizes the construct rather than any of the forces, ideas or elements that form it. It nurtures a mindset that values all and every one that makes up the totality of the construct. We can call this mindset the "construct-truth mindset." When it comes to humanity, people with such a mindset would value the whole of humanity as well as all races that form it.

The philosophy of the construct-truth enables people to see the world from a neutral and scholarly rather than personal and self-interested perspective, which helps people to set up rules that are fair to all players who would in turn support them and be willingly regulated by them. With such rules, all people, including the ones who set the rules, could play fairly and eventually benefit together. Whereas rules that are made out of exclusive-truths often favor particular individuals or groups, which are often despised and disobeyed by other players, leading in the end, to the harm of all for the sake of one.

To go one step further, we might say that there is no absolute truth, in the absolute sense in the world, because there are three parts of the universe, the known, the knowable and the unknowable. Since there is part of the universe that is always unknowable to us, we can never claim any of our discoveries is truly *absolute*. Nevertheless, we should still search for the absolute truth in natural science, which has brought humanity many benefits. However, in the world made of human beings, what we really should do is to create qualified original ideas, because they themselves are to form construct-truth. Indeed, there are no such truths in the world of human beings for us to discover. On the contrary, it is we who must originate qualified ideas to make the construct-truth for ourselves.

Chapter 12
Survival First: A Review of Modern Humanity and a Prospect for Its Future

The modern age is defined by its rebellion against, and eventual replacement of the Middle Ages in the West, a process which is generally considered to have begun with the Renaissance of the fourteenth century. In this sense, humanity entered the modern age more than six hundred years. However, historical judgment on the success of modernity is still waiting. Several fundamental uncertainties concerning the superiority and ultimate desirability of the modern age are indeed yet to be determined.

Many people might take it for granted that the modern age is of course much better than previous ones, as it is obvious the modern age has brought improvements in our basic securities, indeed luxuries, such as improved food and agriculture, superior clothing, much better housing, much better transportation, much better education, and much longer life expectancy on average, to name only a few. However, there are many dark sides of the modern age as well. In fact, they are so dark that some people, while giving it the most serious reflections, might rather live in the Middle Ages than the modern times. The darkest of these dark sides is the scale and threat of destruction that modern age has birthed: environmental pollution and other disasters, technologies, such as nuclear weapons, AI and others, that hold the possibility of extinguishing all humanity. Between the benefits that the modern age has brought about and the survival of humanity itself, it won't take much time for people who have any reasoning power to choose the latter one.

Indeed, the very survival of humanity has been once and again challenged in the modern age, another characteristic unique to modernity. In fact, with the development of modern science and technology, more and more people believe it is just a matter of time before humanity extinguishes itself. Furthermore, terms like "silicon-based life" and "carbon-based life" have been coined and the idea that carbon-based life is to be replaced by silicon-based life is being promoted in the light of the theory of evolution invented by Charles Robert Darwin (1809—1882), which might become more and more popular in the future.

Z. Li and D. Braun, *China Being Led and Leading*,
https://doi.org/10.1007/978-981-96-6803-8_12

It is doubtless that the very survival of humanity is at great risk. It is indeed high time for human beings to take measures to safeguard the very survival of humanity before it is too late.

The Drives for Modernisation, the Problems Caused and the Way Out

There are two major drives for modernisation, an objective and subjective drive. The objective drive might be gathered under the names of modern philosophy, science and technology that started to boom after the Renaissance and have been developing at an accelerated speed, helping to create the material foundations for the modern age.[1] Science and technology, together with the rich materials they help produce, have played a decisive role in promoting humanity into the modern age. They have brought great benefits as well as great risks to humanity. The subjective drive is human desire, or greed sometimes, for more comfort of life, which urges scientists to invent more advanced technologies and manufacturers to produce richer products, sometimes at the cost of the environment or others that are vital for the survival of humanity. It is obvious that human beings' ceaseless pursuit of comfort and material gain in life has often backfired, causing many individual and collective problems, including some that have posed serious threats to all human beings.

Environmental pollution has long been brought to the awareness of the public. In June of 1972, The United Nations held the first world conference on the environment in Stockholm, Sweden, and issued the Stockholm Declaration. However, environmental problems have since become more and more serious, of which the top ten are: (1) the ozone layer depletion, (2) the greenhouse effect and global warming, (3) the intensification of the harm of acid deposition, (4) ecosystem simplification, (5) sharp decline of forests, (6) soil degradation, (7) freshwater crisis, (8) marine environmental pollution, (9) contamination of solid waste, and (10) toxic chemical contamination. (Cao 1995, pp. 86–88.) It is clear that humanity has not yet found an effective way to deal with the environmental problems it has created up till now, and the environment in which humanity can survive continues to erode. Before long, humanity could fall into an environmental situation leading to a point of no return if certain means to thoroughly reverse the trend cannot be found.

The environmental problems of the current scale and anthropogenesis are unique to the modern age, as all of them have been caused by the application of modern science and technology that have developed many means to make products to satisfy the needs of the people, while they have not yet developed efficient means to cope with many of the problems their productivity or products have caused to the environment.

From the technical perspective, the situation could be much improved if the following three remedies could be strictly applied. First, to divide technology into different types according to the potential harm they might cause to the environment,

---

[1] Please refer to Chapter 5.

and only the technology that could completely clean up the waste its products might cause to the environment could be identified as maturated and allowed to be put into use for business. Technologies that have not yet become maturated could only be allowed to be applied for the purpose of experiments. With this principle, many environmental problems could have been avoided from the very beginning. For example, nuclear power plants that have caused huge environmental crises, with the Chernobyl nuclear accident and Fukushima nuclear accident as the most tragic, could never have been allowed to build up.

Second, responsibilities to clean up the waste and destruction their products cause to the environment must be given to those who produce them. For example, factories that produce plastics must be given the responsibilities to clean up the waste of their products, so that the huge problems plastics have caused to the environment could be resolved. In the same vein, factories that produce batteries must be given responsibilities to deal with the problem their products cause to the environment. If factories fail to work out a way to deal with the environmental problems, for instance of plastics bags they produce, then they must voluntarily produce bags from materials of other types, such as paper, cloth or bamboo. Factories that produce batteries, under huge responsibilities, will have to sign contracts with each of their customers to reclaim the batteries after they are used up and probably even have to give each battery they produce a unique number so that they could ask the customer who bought it to pay them back if they are fined because this battery is lost and causes harm to the environment.

Third, public awareness, governmental supervision and international cooperation are indispensable. It is important to make every individual well informed of the consequences of environmental pollution, so that everyone becomes conscientious of the need to protect the environment, and eager to fight against any human activities that harm the environment, either big or small. In comparison with individuals, governments shall take more responsibility in issuing strict laws and regulations to protect the environment, and strictly supervise the development of science and technology, products of the factories, as well as individual activities to make sure the environment is well protected.

In the past, numerous companies of the developed countries made money by causing huge environmental pollution in the developing world and walked away without bearing any of the costs. Even nowadays, certain developed countries are still exporting environmental pollution to developing countries, which should not be allowed. Environmental pollution is a problem for all humanity. The idea that a country would not suffer from the consequences of exporting environmental pollution is short-sighted. In fact, pollution in developing countries will sooner or later bring the same harmful effects to the developed countries, exerted pressure back on the world system. Since developed countries have more capital and better technologies to deal with the waste they produce, it makes sense only from a short-term, profit motivated perspective, rather than a long term, environmentally and humanly responsible, to export degradation to developing countries where the wastes will cause much worse problems.

From a strategic perspective, if we dig a bit deeper into the incentives and motivations of modern industrialized society, we find at least three reasons why the problems in the modern age, environmental pollution included, are particularly difficult to solve.

The first reason is that modern science and technology have their own logic that has been playing the leading role with scientists who have largely lost their own free will and are very much led by the logic. It is often assumed that modern science and technology are under human control, since they were invented and have been developed by human beings. However, it is not the case because modern science and technology have their own working and developing logic, of which the implications have not yet been well studied. Science and technology are material-centred, and scientists follow the rules of the material world. The development of modern science and technology has essentially been to discover the rules of the material world and apply them in technological ways, which is largely decided by the logic of the material world, not the free will of humanity. For example, when chips of 20 nm were invented, it is of the material-centred logic for more precise chips such as chips of 10 nm to be invented until 1 nm, or even smaller. For another example, when goats were first cloned, it was a material-centred logic that drove scientists to clone a monkey, a species that is closer to human beings; and when clone monkeys were produced, it was the same material-centred logic that drove scientists to clone human beings unless the strictest regulation was applied to break the logic.

The second reason is that innovation in science and technology has been ardently encouraged in the modern age by all the major systems of the society, including the political, the economic and the cultural, which bring a lot of power, money and honour to people involved. In the modern age, whoever wins the competition of science and technology wins everything in the fields of military, politics, business and others. European colonists conquered every corner of the world with their weapons invented with modern technology. During the First and the Second World Wars, it was those who had more advanced weapons won at last. Whereas, after the appearance of the nuclear weapons, the law for the side to have more advanced weapons to win ceased to apply, which changed the philosophy of war forever. Since nobody could really win in the traditional way after the invention of nuclear weapons, the benefits of wars drop dramatically. In fact, the ages of wars have gone forever at the strategical level. No nation of reasonable sense would wage a major war in the new age, as it often brings about no profits but huge losses.

Nevertheless, it has always be true that people who have more advanced science and technologies dominate or monopolize world markets, bringing huge wealth to the companies and the nations concerned. Urged by the huge temptation of wealth, it is easy for companies, as well as the governments who are responsible for their supervision, to decide to apply immature technologies and to ignore the problems, either environmental or others, their new technologies might bring about, so long as they hold the promise of profit.

In the cultural field, great honour, with the Noble Prize as the most influential, has been offered to those who make innovations in science and technology. Nowadays, many scientists are eager to make discoveries and innovations to claim their honour without much concern or control of the consequences. Albert Einstein, the greatest scientist in the twentieth century openly expressed his regret at having helped to build nuclear weapons, saying it is "one great mistake in my life." (Einstein 2015, p. 20) It is true that nuclear weapons are one of the most dangerous human inventions. It is indeed true that something must be done to prevent scientists from making terrible inventions, as well as to work out of efficient means to deal with the terrible inventions they have already come up with.

The third reason is that people are often lured to ask for more comfort of life that modern science and technologies are eager to promise, so that people are bought in a certain sense. This has prevented them from effectively stopping problems caused by modern science and technology that are not yet maturated. Indeed, in all forms of modern political systems, leaders are eager to please the public so that they are always ready to satisfy the constantly growing desire of the public to have better lives in order to win their support. As a result, the motivations to regulate the areas of modern science and technology that might lead to the improvement of human comfort are compromised.

In comparison to environmental pollution, the problems caused technologies (such as nuclear weapons, AI and others) that could extinguish all humanity are even more urgent. To effectively deal with such threats, fundamental reviews and reforms of the modern age are necessary, which are to be discussed in the below.

Survival First: The First Law for Humanity in both the Traditional and Modern ages

To ensure the survival of humanity was the first law in both China and the West in traditional ages (Li 2011, pp. 158–164). It is also the first law in the modern age, which, nevertheless, has been ignored to a certain degree and must be highlighted and reconfirmed.

It was never a question as for what to survive before the modern age. However, it is now one of the most important questions to be answered, especially with the advance of biological and AI technologies that are redefining what it means to be human.

With the development of modern science and technology, there might appear human beings whose bodies are very different from the traditional one or, let's name it for the convenience of argumentation, the *classic human beings*. What we mean of the survival of humanity really means the survival of the *classic human beings* who have a natural birth, grow up in the natural way, and live a natural life, like all people in the traditional ages, and almost all people up till now. There must be very strict rules to obey when new technology is to be applied to make human individuals different from the *classic human beings*, if it is to be approved at all.

An individual has to survive and reproduce in a group. The more various groups of people there are, the more chances there are for humanity to survive as a whole.

In the meanwhile, human beings have to survive together with other forms of life, including all sorts of animals and plants. The more various other forms of life are, the more chances one has to survive. Furthermore, we have to survive within a benign ecosystem that can sustain our lives and those other lifeforms. We certainly could not survive in a polluted world that could not sustain a healthy ecosystem. This is why we must protect our world from pollution, protect other forms of life, and protect other groups of people.

In philosophy, some scholars are used to asking ontological questions such as who am I? René Descartes once put forwards the famous argument "I think, therefore I am," (Descartes 1971, p. 103) which could mislead people to conclude that the essence of humanity is *thinking*. With such an understanding of humanity, AI could be said to possess the qualities of humanity since it can process information too. However, it is against the common sense of how we define *classic human beings*. In fact, all attempts to define human beings from the ontological perspective do not apply when we come to the topic of the survival of humanity, as they "always lead to the ignorance of the life of human beings to a certain degree." (Li 2011, p. 158) Our definition for a *classic human being* is as follows: it is whatever makes a complete human being, including the blood and flesh, the knowledge and the skills learned, the cultural values believed in, together with the heritage chain that has led to its birth, the millions of years' evolution that has granted it all the genetic information that has helped us survive for millions of years, and so on. The millions of years that have contributed to the formation of human beings is one of the key factors that define a *classic human being*, which cannot be replaced by any modern, manmade technology.

To give a clear definition of the *classic human beings* is crucial for survival in the modern age when science and technology are developing at an accelerated speed. Both Stephen Hawking and Elon Musk openly warned that AI could cause the extinction of humanity if not well regulated. Stephen Hawkingsaid that AI could be the last milestone of development in the history of humanity, while Elon Musk claimed that AI is even more dangerous than the nuclear weapons. (Wan 2015, p. 15) Some people have been spreading the rumour that the silicon-based life will sooner or later replace carbon-based life in all sorts of social media, without making it clear that what might follow from this initial replacement is the total extinction of humanity, which no reasonable people could have suggested. The very name *silicon-based life* is delusive and misleading, as *silicon-based life* is not *life* at all according to our definition of humanity. Indeed, any so-called *silicon-based life* is just another kind of machine invented by humanity, and will always just be machine.

The boundary of human knowledge is decided by the senses of human beings, such as seeing, hearing, smelling, tasting and touching. The capacity of all human made machines, AI included, is also decided by the same types of human senses, though machines could do better in certain senses. In other words, if there is a dimension that human beings cannot sense, human made machines cannot sense or know it either. It means that all human-made machines are in the same sphere with humanity in terms of knowledge and capability, though they can help human beings to extend their senses to a certain degree.

In contrast to AI, which is a potential threat to the survival of humanity, nuclear weapons have been a real threat for decades, which all human societies have failed to deal with it, helplessly awaiting the possible accidental or intentional use of their destructive capabilities, seeing it in horror to hang above us like a Sword of Damocles. While our memory is still fresh on how close the world was to nuclear war that could have extinguished humanity during the Cuban Missile Crisis, Russia invaded Ukraine in 2022 and openly threatened to use nuclear weapons once again. This put the world into a situation that was even more dangerous, though it seems people are more numb to the threat which makes the world even closer to nuclear disasters.

The fundamental reason for the crises of nuclear weapons to occur and reoccur is the imbalance between the progress of modern science (& technology) and modern cultural philosophy (& social institutions). If we cannot develop mature modern cultural philosophies and social institutions that are capable of fully comprehending and removing the threat of nuclear weapons, crises of nuclear weapons will surely reoccur, making the world more dangerous each time.

It seems that most people have not realised that the world entered into a new phase when nuclear weapons were invented, a new phase in which the most fundamental feature of war had been totally changed: there could be no winner during a war of nuclear weapons. This has really changed the most fundamental philosophy of wars, making wars not attractive anymore. After the invention of the nuclear weapons, no major wars like those of the First and the Second World Wars have been fought. After the invention of nuclear weapons, wars between nuclear powers are no longer the final solution for disputes. This really means that the ages in which wars played the most fundamental role have perhaps gone forever, and a new phase of the modern age has come, in which people must find ways to settle disputes via peaceful means.

The wars fought after the invention of nuclear weapons were caused because the mindset fostered during the traditional ages are still alive, while modern cultural philosophy and social institutions of the new phase had not yet been well developed. Undoubtedly, any war or threat of war will urge nations who do not have nuclear weapons to seek cover of nuclear weapons either by producing them or by joining a military ally that does. In due course, there will be no nations who are not covered by nuclear protection. By then, even a war of a small scale could lead to nuclear disasters, which is surely not what we want to see in the future.

To avoid all the crises of both AI, nuclear wars or the like, and to ensure the survival of humanity, it is most necessary to develop modern cultural philosophy and social institutions, for which something can be done in the following three perspectives.

The first and the most important is to promote love rather than hatred between different groups of people. So long as there are different individuals, there are conflicts of interest, hence disputes, arguments and fights. However, we can handle them in all different ways. The best model to handle such conflicts is that among family members, in which love and compromises are promoted. As a result, peaceful methods to settle disputes could always be found. In fact, if we do believe human beings are a family, we could always apply the same family model to settle disputes between different groups and nations.

Needless to say, it is extremely hard for some people to put aside their biases and hatred against others that are planted in their hearts by all the religious, cultural or political institutions that have shaped their views. However, this must be done, no matter how hard it is, so long as human beings hope to ensure their survival in the new phase of the modern age. If people truly care for the future of humanity, as well as their own future, and are determined to promote love instead of hatred, they will be successful. When they succeed, they will find that it is in fact not difficult to live harmoniously with other peoples at all, and the rewards will be tremendous and forever-lasting.

Second, it is necessary to build up an international legal system and corresponding enforcement forces that treat every individual of human beings equally and fairly. So far, some developed nations have designed a legal system that treats each of their citizens equally and fairly, so that the legal system is recognised by all their citizens and so the nations are regulated. In the new phase of the modern age, such a legal system that treats each human individual equally must be invented and enforced, in order to bring a world in which all people trust the legal system, which will lead to a world ruled by law, not men. Indeed, we do not lack the wisdom. What we lack is the good will to treat other people as our family members.

Third, international trade, as well as other means of international cooperation, must be encouraged. After the Second World War, led by the United States, the West built up a system in which international trade has prospered, which has effectively enhanced interdependence between Western nations, so that it is of the greatest interest for all members of the Western nations to keep the system and of no interest for any member to start a war against the system. There had been endless bloody wars of large and small scale among Western nations before and during the Second World War. However, with the prosperity of international trade and other communications, the West has remained at peace. In the same vein, if international trade and other means of communications prosper among all nations in the world, all nations involved will pursue peace rather than war.

It is a great pity that all nations, including the US, are still viewing the world in the mindset that was formed in the past, so that they are taking their own national interest as the highest law. As a result, the world is not as prosperous for all as it could be. In the new phase of the modern age when the very survival of humanity is at great peril, the most important task for all nations is to work together to ensure the very survival of humanity, without which the interest of a nation is meaningless. In the new phase of the modern age, the survival of all humanity should be taken as the first law above anything else, even above the national interest of a nation.

If the US had put the survival of all humanity above its national interest, it should have taken the opportunity to negotiate with Russia to destroy all or most of the nuclear weapons that could extinguish humanity after the Soviet Union dissolved in the 1990s. Unfortunately, the US celebrated as a winner, in the mindset formed in the past, and so it missed this great opportunity and allowed the crisis of nuclear wars to come back again after the start of the Russia-Ukraine war.

It is self-evident that no one can survive if humanity perishes as a whole. It is also self-evident that all technical, economic and other developments are worthless if the very survival of humanity cannot be assured. Therefore, it is self-evident that the very survival of humanity must be taken as the first law to guide human beings in the new phase of the modern age.

The Root Faith of Life and the Modern Cultural Value System Based upon It

In comparison to the survival of humanity that is the highest law for human societies, the most fundamental faith of human beings is faith in life, not only the life of humanity, but also lives of other forms that sustain us.

Faith in life is the root faith for all other faiths and values. Indeed, it is the very root for humanity, and is carried in the genes of all human beings and human cultures, which span traditional and modern ages(Li 2011, pp. 158–164). As soon as a life is born, it instinctively strives to live on, which vividly expresses that faith in life and life persistence is of the nature of all forms of life. It is out of the same faith in life that human beings have invented and developed all sorts of cultures, which implies that faith in life is the very root for all human cultures. This is why all human thoughts and acts are ultimately evaluated according to their contribution to human lives.

The root faith of life is not only physical life, but also social life and conceptual life.[2] The root faith of life is not only one's own life, but also the reproduction of one's life, so as to carry on humanity. Every individual is going to die sooner or later. However, human beings as a whole must survive if humanity is to continue on. The root faith of life is not only of human lives, but also of lives of other forms on earth, without which human beings could not survive.

The root faith of life does not forbid human beings to kill other forms of lives. However, it does require human beings to cherish, respect and protect them at the strategic level, and take the lives of other forms only when it is the most necessary. For example, it is fine to kill an animal to feed human beings. However, it is immoral to kill an animal without any sound reason, or to kill an animal without making full use of its sacrifice, or to kill an animal without thanking it, or to kill an animal that is looking after its children, or to kill an animal that is of an endangered species, and so on.

The root faith of life demands that human beings do not take the lives of other human beings except for certain extreme situations, such as for self-defence. The root faith of life believes any war is the tragic result of human beings' stupidity so that it does not approve any war that is designed to kill human beings. In the traditional ages, fights between villagers of different family tribes in China, for example, were often fought causing many deaths. However, there are not such fights any more in contemporary China. Hopefully, the same change could happen all over the world so that there will not be any war in the future when the world gets more organised.

---

[2] Please refer to Chapter 8: "Who Am I?": Establishment of Modern Culture on Basis of Three Forms of Personal Life.

Despite the fact that there are still wars in the world, the root faith of life has been accepted to a certain degree even in military operations, as evidenced by the fact that it has been widely accepted that a surrendered soldier must not be killed. This demonstrates that the root faith of life has already been accepted by people before they realise it.

All human-invented cultural, religious, and political institutions, and the like were meant to serve lives. In fact, all well-known institutions have made great contributions to the groups of people they have served. However, due to the fact that some institutions have not observed the root faith of life closely, they have deviated from whatever they were meant to serve, and caused a lot of problems, even disasters. For example, Confucianism was invented to help people who followed it, and it indeed helped its followers. However, it sometimes deviated from the root faith of life and caused a lot of harm or even deaths, which is why Lu Xun fiercely attacked it in his classic work, "A Madman's Diary," as of a nature that eats people. (Lu 1918, pp. 414–424.) It is the same with Christianity, which was meant to be help people who followed it, and it indeed helped its followers. However, Christianity sometimes deviated from the root faith of life and caused a lot of harm and death to its own people as well as others.

In the modern age, many traditional institutions, with Confucianism and Christianity included, were reformed with the root faith of life as the leading principle, so that their followers in the modern age might enjoy a much happier life. However, there are still many problems that need to be tackled. For example, there are still wars, though they are less frequent, and less deadly. For another example, people of different racial, religious, cultural, national or political groups may distrust or even hate each other, which has caused many tragic events, for example the 9/11 Terrorist Attacks on the US.

Besides the problems caused by flaws in all sorts of institutions, the development of modern science and technology have brought and are bringing much more serious problems, as discussed above. Modern science and technology were invented to help life and are driven forwards for the same purpose. Therefore, it is a matter of course for us to demand they observe the root faith of life, so that any development of science and technology which runs athwart the root faith of life should be either suspended or banned upon consideration of its potential harm to humanity.

To deal with all the problems, either those caused by flaws of institutions or those caused by the development of modern science and technology, it is always important to highlight the root faith of life that will help us to constantly correct many mistakes occurred or to occur.

With the root faith of life as the criterion, we can review all institutions set up and adjust or reform them accordingly. All existing cultural, religious and political institutions share a common problem: they divide people into us vs others, which only care for "us," not the "others." The root faith of life holds the view that all human beings are of the same life and the same family, and proposes family love for all human beings.[3] With this principle, all sorts of institutions could be adjusted or reformed, which might significantly enhance love and harmony between different groups of people.

With the root faith of life as foundation, we can establish a system of modern values to guide our behaviour in the new phase of the modern age. There are many values that support the root faith of life, such as to be kind to others, to be self-disciplined, to be humble and so on. To promote such values in the modern age is critical, and ought to form a core component of all education, which should be conceived of as means of preparing a person ready for the modern way of life. Without observing key modern values, one could neither enjoy the modern way of life nor to make any contribution to it. Ideas that are directly against the root faith of life, such as those of fascists, racists, extremists and the like should be carefully guarded and fought against with determination.

With the root faith of life as guide, we can regulate the relationship between the good and the true, letting the good lead the true.[4] Even the system of knowledge should be readjusted so that the subject of ethics is placed at the core. (Li 2010, pp. 968–974) A person should be a person of virtues first, and then be a person of other businesses.

Conclusion: A Prospect for the Maturated Modern age of Humanity

Since entering the modern age, humanity has benefited greatly from the development of modern science and technology. Meanwhile, humanity has also had to face many new challenges, especially in the wake of the invention of the nuclear weapons, which made the extinction of humanity within the reach of a small group of people who should have never been given such power. It is in fact a milestone to mark that humanity had entered a completely new phase of the modern age, though not many people have realised it.[5]

In this new phase, humanity is facing problems it had never encountered in traditional ages. Nor could we hope to get the problems well tackled with the cultural, religious or political institutions that were formed in the past. Therefore, we have to apply a forward-minded vision, and readjust or even rebuild all systems to absolutely ensure the survival of humanity, i.e., to absolutely defuse all perils caused by

[3] Please refer to Chapter 7: "Where Am I Going?": Reconfirmation of the Fundamental Cultural Faith in Life of Humanity and Its Survival.

[4] Please refer to: Zhimin Li. 2011. *Zhishi de Shanyuzhen* [The Good and the True of Knowledge]. Beijing: Renmin Chubanshe.

[5] Ulrich Beck said: "since the middle of this century the social institutions of industrial society have been confronted with the historically unprecedented possibility of the destruction through decision-making of all life on this planet. This distinguishes our epoch not only from the early phase of the industrial revolution, but also from all other cultures and social forms, no matter how diverse and contradictory these may have been in detail" (Beck 1992, p. 101).

humanity itself, which have threatened the very survival of humanity. This is what we should do, and what we can do. After it is done, we will be able to enter into the *maturated* phase of the modern age.

Human beings have been working hard to develop modern science and technology to serve the interests of human beings. However, it is evident that all such developments are pointless if human beings are to be killed by whatever they have developed, or are developing. It is therefore without doubt that the survival of the human being should always be placed over the development of modern science and technology, as well as the comfort their inventions might provide. It is high time for human societies to accept survival as the first law in the new phase of the modern age, with which modern civilizations are to be critically reviewed, adjusted or rebuilt on basis of the root faith of life as foundation. For example, nuclear weapons owned by all nations should be totally or largely destroyed, so that they cease to pose a threat to humanity as whole; strict rules should be imposed to govern the development and application of AI, and so on; the modern economical system, with its economic mechanism that has provided the impetus for development of modern science and technology needs to be reviewed, qualified and adjusted. Indeed, in the new phase of the modern age, a new human civilization is to be born to secure the survival of humanity and promote its progress.

Works Cited

Beck, Ulrich. 1992. From Industrial Society to the Risk Society: Questions of Survival, Social Structure and Ecological Enlightenment. *Theory, Culture & Society* 9: 97–123.

Cao, Lei. 1995. Quanqiu Shida Huanjing Wenti [A Review on Ten Problems of the Global Environment]. *Huanjing Kexue* [Environmental Science] 4 (16): 86–88.

Descartes, Rene. 1971. *Discourse on Method and the Meditations*, trans. F. E. Sutchiffe. London: Penguin Books.

Einstein, Albert. 2015. *Bite-Size Einstein: Quotations on Just About Everything from the Greatest Mind of the Twentieth Century*. New York: St. Martin's Press.

Li, Zhimin. 2010. Yi Lunlixueshuo Weilinghunde Duoweilitixing Zhishitixi Jiangou [Constructing a Multi-Dimensional Model of Knowledge Classification System on Basis of Taking Morality as Its Spirit, Studies in Science of Science]. *Kexuexue Yanjiu* [Studies in Science of Science] 7: 968–974.

Li, Zhimin. 2011. Shengzhixin: Guanxi Jingweizhongde Wenhua Qiankun [Faith in Life: Culture in Relationships]. *Lanzhou Xuekan* [Lanzhou Academic Journal] 11: 158–164.

Lu, Xun. 1918. Kuangren Riji [A Madman's Diary]. *Xin Qingnian* [New Youth] 4 (5): 414–424.

Wan, Cheng. 2015. Zhongjiezhe? Huojinyu Masike doudui Rengongzhineng Biaoshi Danyou [Terminator? Both Stephen William Hawking and Elon Reeve Musk Expressed Concerns of AI]. *Huadong Keji* [East China Science & Technology] 3: 15.

Coda

For the Mature Modern Age

We learn lessons from history, but we do not live in it!
We move forwards!
When we move forwards, we need a future-oriented mind to guide us.

With a future-oriented perspective, we find that the real enemy of all humanity in the modern age is the abuse of modern science and technology that has pushed humanity to the very brink of extinction. To ensure our very survival, humanity must unite as a family (*we really are*) to defeat the enemy that has been built up, ironically, by our own greed, stupidity and lack of self-discipline. When we succeed, humanity will have entered the mature phase of the modern age, assured of its very survival and able to truly enjoy the benefits brought about by modern science and technology.

It is a task that demands a lot of effort and time. However, as individuals, we can start at any minute: we can observe modern cultural values and live a virtuous life according to standards of the mature modern age that will come sooner or later.

Z. Li and D. Braun, *China Being Led and Leading*,
https://doi.org/10.1007/978-981-96-6803-8

Index

A
Abstract Social Being, 82, 91, 98
Abstract thinking, 115–118
Acid deposition, 152
Act of Supremacy (1536), 67
A Dream of Red Mansions, 33, 131
AI (Artificial Intelligence), 94, 95, 98, 151, 155–157, 162
American Declaration of Independence, 84
American political system, 134
Ancient Greece, 62, 65
Andrews, Bruce, 53
A Poem-Selection of the Most Popular Chinese Poets in the 1990s, 53
Aristotle, 62, 66, 116, 122
Arnheim, Rudolf, 116, 119
Artistic works, 121, 131
Arts and humanities, 124, 140, 141, 143, 145, 149
Assisted Reproductive Technology (ART), 93
Attempting Poems, 31, 34, 48, 49
Authoritarian, 108

B
Bacon, Francis, 14, 69, 70
Bai, Hua, 53
Baihua, 46, 52
Bai, Juyi, 44, 46
Basic consensus, 141–143, 145, 146
Basic securities, 151
Baylor, A., 115
Beauty-ugliness connection, 124
Beck, Ulrich, 161
Benedict XIV, 14
Bernstein, Charles, 8, 53, 131, 133, 134
Bian, Zhilin, 50
Big Bang theory, 80, 81
Body thinking, 119, 120, 123, 125, 126
Book of Etiquettes, 107
Bruno, Giordano, 65
Budding period, 141, 143, 145, 146
Byron, Lord, 53

C
Calvinism, 68
Cambridge, 3, 7–9, 13, 18, 19, 21, 22
Cao, Yu, 30, 35, 36
Carbon-based life, 95, 151, 156
Cathay, 18, 44
Chen, Duxiu, 30, 35, 46
Chernobyl nuclear accident, 153
Chinese Couplets, 44
Chinese Language-Poetry Group Che Qianzi, 54
Chinese literature, 17, 20, 24, 30, 31, 33–35, 37, 44–48, 50, 51, 56
Christianity, 14, 61–67, 71, 72, 78, 83, 86, 92, 93, 96, 136, 146, 148, 160
City-states, 61, 62
Civil education, 81
Civil War, 29, 63
Classic human beings, 155, 156
Classic of Filial Piety, 76

Z. Li and D. Braun, *China Being Led and Leading*,
https://doi.org/10.1007/978-981-96-6803-8

Cloning, 94
Cloning technology, 94
Cognitive construct, 125
Common Morality, 60–65, 70–72
Compass, 23, 43, 69, 70, 103
Concept thinking, 118–126
Conceptual self, 101, 102, 104–108
Conceptual thinking, 125
Confucianism, 36, 45, 59–61, 70–72, 83, 86, 90–93, 107–110, 112, 118, 135, 139, 145, 146, 160
Confucian scholars, 35, 76, 77, 91, 107
Confucius, 35, 59–61, 77, 83, 86, 87, 89–91, 93, 107, 109
Construct-truth, 140–145, 148, 149
Construct-truth mindset, 149
Copernicus, Nicholas, 65
Counterrevolutionary Camp, 30
Critique of Practical Reason, 69
Cuban Missile Crisis, 157
Cultural codes, 107
Cultural communication, 126, 134
Cultural education, 105
Cultural equality, 85
Cultural faith, 78, 79, 83, 86, 89, 92, 94–99, 105, 112, 161
Cultural faiths and values, 101, 105, 112
Cultural institutions, 29, 38, 83, 85, 86
Cultural philosophy, 87, 92, 94, 95, 111, 157
Cultural Revolution, 33, 36, 37, 40, 50, 51
Cultural rights, 84–86, 96, 97
Cultural rights and obligations, 76, 78, 83, 102
Cultural transformation, 37, 140
Cultural values, 8, 15, 21, 34, 35, 37, 39, 75, 76, 81, 83, 84, 86, 87, 89, 96, 98, 104, 109, 110, 139, 156
Cultural value system, 36, 85, 86
Cultural vanguard, 45

D
Dachun, Hei, 53
Dai, Wangshu, 50
Darwin, Charles, 2, 3, 79
Death, 2, 23, 24, 27–29, 63, 65, 77, 80, 84–86, 89–93, 96, 97, 104, 106, 108, 111, 142, 146, 159, 160
Decline of forests, 152
de Saussure, Ferdinand, 122, 129
Descartes, René, 68, 69, 71, 115, 116, 125, 156
Diachronic construct-truth, 145
Doctrine of fate, 76
Dostoevsky, Fyodor, 80

E
Ecclesiastical power, 78
Economic growth, 38
Ecosystem simplification, 152
Edict of Milan, 64
Educational opportunities, 106
Educational system in China, 147
"Eight Nos", 48
Einstein, Albert, 124, 143, 155
Eliot, T.S., 44, 130–133
Emancipation of slaves, 63
Emerson, Ralph Waldo, 3, 136, 137
Emperor-centered, 108
Emperor Wudi, 60, 61
Emperor Yongzheng, 14
Empson, William, 18, 21, 22, 24
Endless Wars and Sufferings, 27
Environmental pollution, 151–155
Equality, 9, 15, 25, 37, 62, 75, 85–87, 94, 101, 110
European Romanticism, 52
Evolution, 82, 94, 101, 119, 122, 124, 126, 156
Evolutionary history, 82, 126
Evolutionary perspective, 116
Exclusive truth, 140–143, 149
Exclusive-truth mindset, 149

F
Family construct-truth, 144
Family values, 36
Fan, Xiao, 115, 117, 118
"Farewell Again to Cambridge", 18–21, 24
Fascists, racists, extremists, 161
"Fear of Flipping", 131, 133, 134
Filial obedience, 85, 110
Filial piety, 75, 77, 78, 84, 85, 87
Final judgment, 92
Flexibility of human beings, 143
"Flowers of Two Persons", 54
Four Tones of Chinese language, 56
Freedom of speech, 146, 147
Freedoms of speech and press, 105

Free love and marriage, 36, 39, 134
Freshwater crisis, 152
Friedrich Schlick, 124
Fukushima nuclear accident, 153

G
Gerard of Cremona, 66
Gernet, Jacques, 14, 45, 136
Globalization, 43
Global warming, 98, 152
Good-evil connection, 124
"Good Morning", 49
Governmental supervision, 153
Greek culture, 61
Greenhouse effect, 152
Gu, Cheng, 52
Gunpowder, 14, 23, 43, 66, 69, 70, 103
Guo, Moruo, 30, 33, 35, 49

H
Haizi, 53
Han Dynasty, 28, 55, 106
Hawking, Stephen, 80, 81, 156
Hemingway, Ernest, 53
Henry VIII, 67
Hereditary obligations, 84
Hierarchy of Needs, 104
Historical judgment, 151
Homer, 62, 65
Huang, Yunte, 131–133
Huang Zunxuan, 56
Human divinity, 79, 83
Humanity, 3, 8, 22–24, 51, 56, 76–84, 86, 87, 89, 91, 94–99, 101, 106, 113, 136, 143–145, 149, 151–163
Human senses, 116, 117, 156
Human survival, 116, 126
Hu, Shi, 17, 18, 30–34, 36, 45–49, 51

I
Ideological development, 101
"I Have a Dream" speech, 63
Image-Encoding, 120
Image thinking, 115–126
Imagism, 18, 48, 52
Immortality, 91
Imperial system, 27–29, 108, 109, 111, 112, 139, 145, 146, 148
Inalienable rights, 112
Independent thinking, 105, 108
Individual independence, 110
Industrialization, 38, 39, 136
Information connections, 123–125
Information process, 116–120, 123
Innovation, 29, 50, 109, 124, 125, 130, 132, 134, 135, 147, 154
Innovation in science and technology, 154
Inspiration, 28, 49, 70, 133, 134
Inspirational thinking, 115
Instinctive knowledge, 125
Instinctive response, 119
International community construct-truth, 144, 145
International cooperation, 153, 158
International poetics, 130, 134
I Sang for Happiness, 50

J
Ji Xian, 17
Judgment-making, 125

K
Kant, Immanuel, 68, 69, 71
Keller, Helen, 130, 136
King, Martin Luther (Jr.), 63
Knowledge, 9, 14, 18, 48, 68, 69, 81, 83, 86, 87, 106, 107, 109, 111, 116, 122–125, 132, 134–136, 140, 141, 143–147, 156, 161
Kublai Khan, 65
Kuomintang, 29

L
L=A=N=G=U=A=G=E, 53
Langage, 129
Language School, 132, 133
Langue, 129
Leaves of Grass, 53
Legal rights, 112, 147
Liang, Xiaoming, 53
Li Bai, 44, 135, 141, 145
Life chain of humanity, 84, 85
Light Rain, 49, 50
Li Jinfa, 49, 50
Lincoln, Abraham, 63
Lin, Dongwei, 53
Linguistic connection, 123
Linguistic sign, 123, 129, 130
Linking, 132, 133, 137
Lin Yutang, 19
Literary and cultural mentality, 27, 30, 37

Literary language (Wenyan), 31–33, 46–48, 52
Literary Revolution, 17, 30–33, 35, 36, 40, 48, 51
Literary studies, 9, 126
Li Yu, 54
Li, Zhimin, 3, 7–9, 17, 19, 24, 30, 31, 37, 80, 81, 85, 92, 101, 132, 133, 135, 136, 140, 155, 156, 159, 161
Logical-rational thinking, 124
Logical thinking, 115, 116, 118, 124
Longevity, 96
Lowell, Amy, 48
Lutheranism, 68
Luther, Martin, 67
Lu, Xun, 30, 32–34, 36, 37, 39, 51, 160
Lyric, 53, 134, 135
Lyric poems, 44

M
Ma, Li, 53
Mao, Zedong, 47, 55
Marine environmental pollution, 152
Marriage of Xiao, 135
Marx, Karl, 102
Maslow, Abraham, 104
Master-disciple relationships, 86
Material-centred logic, 154
Material wealth, 95, 98
May Fourth Movement, 15, 19, 24, 30, 46, 49, 86, 109, 134, 139
Medieval Europe, 65, 66
Middle ages, 78, 148, 151
Milton, John, 55
Misty Poetry, 51–53, 55
Modern age, 7, 8, 28, 35, 38, 40, 43, 45, 47, 71, 81–87, 89, 94, 96–99, 101, 103, 106, 107, 110–113, 118, 122, 136, 140, 142, 143, 146–149, 151, 152, 154–163
Modern Chinese novels, 34
Modern Chinese plays, 34
Modern cultural and political systems, 37, 112, 113
Modern cultural faith, 112, 124
Modern cultural faiths and values, 112
Modern cultural norms, 75
Modern cultural system, 109–111
Modern cultural values, 9–11, 37, 81, 83, 86, 87, 95, 97, 110, 159, 163
Modern economical system, 162
Modern history of China, 27, 60
Modern industrialized society, 154
Modernists, 17, 44
Modernity, 15, 27, 38, 78, 86, 110, 113, 132, 151
Modernization, 17, 37, 40, 48, 55, 71, 86, 139, 140
Modernization of China, 40
Modern medicine, 93–95, 97
Modern nation, 29, 113
Modern philosophy, 40, 80, 152
Modern poetry, 7, 35, 131–133
Modern political system, 29, 111, 112, 155
Modern science, 3, 23, 24, 37, 38, 65, 69–71, 79–83, 86, 92–95, 98, 99, 112, 151, 152, 154, 155, 157, 160–163
Modern science and faith, 87
Modern societies, 39, 66, 83, 85, 86, 94–96, 98, 112, 146
Modern technology, 38, 70, 112, 154
Modern values, 37, 39, 40, 161
Modern Xiao Dao, 85
Monasteries, 66
Moral and immoral, 99
Moral inefficiency, 110
Morality, 24, 60–62, 64, 69–71
Mosaic covenant, 83
Moses, 64, 80, 83
Mo, Yaping, 53
Musk, Elon, 156

N
Narrative works, 134
Natural connection, 123
Natural death, 97
Natural disasters, 108, 145
Natural evolution, 98
Natural order, 97
Natural science, 16, 17, 43, 124, 140, 141, 143, 149
Needham, Joseph, 14, 23, 43, 63, 70, 71, 103
Nero's bloody repression, 64
New Chinese Poetry, 17, 19, 20, 22, 31, 45, 47–56
New human civilization, 162
New Youth, 30, 46
Nie, Pei, 53

9/11 Terrorist Attacks, 160
Ninety-Five Theses (1517), 67
Nuclear power, 94, 153, 157
Nuclear weapons, 94–96, 98, 151, 154–158, 161, 162
Nurturers, 84
Nüwa creating humanity, 76, 79, 87

O
Obama, Barack, 63, 142
Objective correlative, 130
Opium War, 15, 45, 46, 139
Origin of life, 79–81
'Over the Roofs', 17
Ozone layer depletion, 152

P
Panoramic thinking, 123, 124
Parent-child relationships, 75–77, 84–87
Perloff, Marjorie, 133
Personal cultivation, 9, 101
Personal life, 102
Philosophy of rational divinity, 87
Philosophy of thinking, 115, 120, 124
Physical exercises, 125
Physical self, 101–106
Piaget, Jean, 121
Pius VI, 14
Plato, 62, 68, 69, 124
Polanyi, Michael, 116
Political agreements, 99
Political construct, 148
Political rights, 97
Political system, 28, 67, 85, 106, 111
Pound, Ezra, 18, 24, 44, 48, 131, 135
Pragmatic approach, 90
Printing, 14, 23, 43, 66, 69, 70, 103
Procedural consensus, 141–143, 146, 148
Protestant Reformation, 67
Public awareness, 153
Public opinion, 104
Pushkin, Alexander, 53

Q
Qian, Xuesen, 115, 124
Qi Baishi, 56
Qing Dynasty, 28, 33, 38, 45, 77, 109
Qualified original ideas, 144–149
Quality of life, 97, 106, 107

R
Rational divinity, 87
Rationality, 87, 89, 92, 118, 124
Rational theist, 87
Rational thinking, 116, 118, 119, 124
Reform and Opening-Up Policy, 4, 8, 27, 37, 38
Religion, 2, 16, 61, 62, 64, 65, 67–69, 71, 72, 78, 80, 81, 83, 92, 93, 103, 146
Renaissance, 45, 48, 66–68, 151, 152
Reproduction, 84, 96, 159
Republic of China, 28, 29, 134
Revolutionary Camp, 30–36
Ricci, Matteo, 14, 22
Richards, I.A., 47, 56
Roman Catholic Church, 67
Roman Empire, 64–67
Rule by virtue, 61, 77
Rulemaking, 99
Rule of law, 9, 81, 111, 147
Russia-Ukraine war, 158

S
Sanxiangu, 135, 136
Schema theory, 123
Science and Democracy, 30, 139
Science and technology, 3, 24, 37, 38, 43, 69–71, 151–156, 160–163
Scientific problems, 124
Second World War, 29, 154, 157, 158
Self-actualization, 104
Self-subjugation, 107
Seniors, 2, 3, 5–9, 38, 96, 97, 107, 121
Sensory adaptation, 119
Shakespeare, William, 22, 53
Shenshen Poetry Exhibition of 1986, 52
Shenzhen Youth Newspaper, 53
"Shi Guai", 50
Shu, Ting, 52
Signified, 129, 130, 136
Signifier, 129, 130, 135, 136
Silicon-based life, 95, 151, 156
Sino-Japanese War, 15, 17, 38, 139
Slave systems, 62
Social and natural fields, 112
Social being, 16, 90, 91, 102, 122
Social consensus, 104
Social construction, 101
Social development, 60, 92, 146
Social disputes, 98
Social flourishing, 110
Social institutions, 81, 157, 161

Socialism with Chinese Characteristics, 134
Social mobility, 106
Social norms, 80, 104
Social order, 70, 107, 109, 110
Social organization construct-truth, 144
Social responsibilities, 60
Social sciences, 43, 124, 140, 141, 143, 145, 149
Social self, 101–106
Social stability and security, 108
Socrates, 68
Soil degradation, 152
Solid waste contamination, 152
Songs from the South, 47
Sound-image, 44
Southern economy, 62, 63
Spiritual life, 90, 91, 98
Spiritual wealth, 96
Stowe, Harriet Beecher, 63
Structuralist theories, 140
Sun, Yat-sen, 6, 28, 29, 134
Survival, 82, 89, 90, 94–99, 101, 106, 119, 126, 151, 152, 155–159, 161–163
Synchronic construct-truth, 145

T
Tagore, Rabindranath, 53
Taiping Heavenly Kingdom Movement, 28
Tang Dynasty period, 44
Taoism, 61, 71, 90, 118
Teacher-student relationships, 86
Teasdale, Sara, 17
Ten Commandments, 64, 80, 83
The *Bible*, 13, 64, 65, 67, 68, 71, 77–80, 82, 83, 86, 91, 92
The Chinese Written Character as a Medium for Poetry, 135
The Crusades, 65
The Fortunate Union (《好逑传》), 16, 44
The Goddesses, 49, 50, 56
The Long-Term Visitor and Hard Times, 50, 51
The Old Man and the Sea, 53
The *Old Testament*, 64, 80
The Orphan of China (《赵氏孤儿》), 16
Theory of evolution, 2, 78–80, 151
"The River-Merchant's Wife: A Letter", 44
The Selections of the Jottings of Contemporary Chinese Poets, 53
Three Obediences and Four Virtues, 110
Tian (Heaven or God), 71, 82, 84, 93
Toxic chemical contamination, 152
Traditional age, 29, 39, 55, 89, 96, 98, 146, 155, 157, 159, 161
Traditional China, 23, 28, 31, 33, 35, 36, 54, 70, 77–79, 85, 86, 106–112, 139, 145, 148
Traditional Chinese culture, 15, 36, 38, 45, 61, 75–77, 86, 97, 122, 123
Traditional Chinese poetry, 18, 31, 44, 46–49, 54, 56, 134
Traditional Confucianism, 85, 89, 96, 110–112, 148
Traditional Western culture, 77, 78
Translation, 14, 16–20, 24, 32, 33, 43, 51, 66, 67, 75, 133, 135, 140
Transplantation, 17, 134, 136
True-false connection, 124
Turbulent Transformation, 27
Twenty-Four Filial Exemplars, 77

U
Uncle Tom's Cabin, 63
United States of America (USA)/United States (US), 6, 8, 18, 29, 30, 46, 48, 53, 62, 63, 71, 83, 93, 113, 142, 148, 158, 160
Urbanization, 38, 136

V
Vernacular Chinese, 31–34, 46–48, 52
Virtues, 62, 64, 77, 90, 91, 110, 135, 148, 161
Voltaire, 13, 16
von Goethe, Johann Wolfgang, 16, 43

W
Wang, Duqing, 50
Wang, Guowei, 47
Wang, Jiaxin, 53
Wang, Zuoliang, 22
Wen, Yiduo, 49, 56
Western ideologies, 30, 45, 46, 134, 139
Westernization Movement, 38
Western poetry, 17–19, 45, 52, 54–56, 134
Western powers, 7, 17, 109, 139
Western Symbolism, 50
Whitehead, Alfred North, 68
Whitman, Walt, 53
Wittgenstein, Ludwig, 65, 71
Wordsworth, William, 134
World War I, 135

X
Xiaodao (孝道), 75
Xi Chuan, 53
Xie, Ming, 135
Xin Hai Revolution, 109, 139
Xiyu, 56
Xu, Guangqi, 14
Xu, Zhimo, 18, 19, 24, 36

Y
Yi, Hai, 53
Yuan Shikai, 28, 29
Yue, Daiyun, 135, 136

Z
Zhang Ziqing, 53, 131, 132
Zhu, Xi, 107

GPSR Compliance

The European Union's (EU) General Product Safety Regulation (GPSR) is a set of rules that requires consumer products to be safe and our obligations to ensure this.

If you have any concerns about our products, you can contact us on ProductSafety@springernature.com

In case Publisher is established outside the EU, the EU authorized representative is:

Springer Nature Customer Service Center GmbH
Europaplatz 3
69115 Heidelberg, Germany

Batch number: 10725798

Printed by Printforce, the Netherlands